SCIENCE
AND
FAITH

SCIENCE
AND
FAITH

*Navigating the Labyrinths
& Connecting the Bridge*

Authored By,
Dr. Khaled Ghandour

**WINGS
PUBLICATION
INTERNATIONAL**
WE GIVE WINGS TO YOUR DREAM!

Disclaimer

WINGS PUBLICATION INTERNATIONAL
WE GIVE WINGS TO YOUR DREAM!

Registered Office- 907-Sneh Nagar, Sapna Sangeeta Road, Agrasen Square, Indore – 452001 (M.P.), India

Website: http://www.wingspublication.com

Email: mybook@wingspublication.com

First Published by WINGS PUBLICATION 2023

Copyright © Dr. Khaled Ghandour

Title : SCIENCE AND FAITH

Price : $ 25 I AED 80

All Rights Reserved.

ISBN : 978-81-19223-84-8

LIMITS OF LIABILITY/DISCLAIMER OF WARRANTY

Image of the Front Cover

This was an early attempt to depict consciousness in a geometrical figure, drawn by Benjamin Betts, and contained within a book written by: Louisa S. Cook, titled: **"Geometrical psychology, or, The science of representation an Abstract of the Theories and Diagrams of B. W. Betts",** published by; G. Redway, London in 1887. The book currently is available for download online as an open-source in the "Internet Archive"[1] and the "Public Domain Review"[2] websites.

Benjamin Betts was a New Zealand mathematician and spiritualist who, during the 1880s, attempted to quantify and model human consciousness in geometric terms. He believed that by visualizing the abstract concepts and states of mind, one could better understand human consciousness.

Betts developed a series of figures, which he called "geometrical psychology" diagrams, representing the evolution of human consciousness. These took the form of intricate geometric designs and were intended to illustrate the various states of consciousness, from the simplest form of existence to the most complex.

---

[1] Digitized by the Internet Archive in 2010 with funding from Open Knowledge Commons and Harvard Medical School. Openlibrary_edition OL24404552M.

https://archive.org/details/geometricalpsych00cook/page/n3/mode/2up

[2] The Public Domain Review is an online journal and not-for-profit project dedicated to the exploration of curious and compelling works from the history of art, literature, and ideas.

The website was established in 2011 and focuses on works now fallen into the public domain, where commons out-of-copyright material is free for everyone to enjoy, share, and build upon without restriction.

https://publicdomainreview.org/collection/geometrical-psychology-or-the-science-of-representation-1887/

While Betts' work never gained widespread acceptance during his time and remains relatively obscure, it represents an early and unique attempt to merge spiritual concepts with mathematical and geometric representation. His work can be seen as a precursor to some of the more abstract art movements of the 20th century, which also sought to represent the inner workings of the mind.

He writes: *The impulses of the emotional life, become positive, and the activity which determines them negative in the Onde. The progressive circles of expansion proceed according to an accelerating ratio as before, outwards for the Ond and inwards for the Onde.*

This image is labeled as figure 9 in the book and is referred to as the Ond. It starts from a center of possibility and extends itself ad infinitum into objectivity, and is meant to depict the disproportion between the ideal and the real, and this causes the bend in the figure and how it forces itself upon the consciousness.

As consciousness is the core of human deliberations and beliefs, it was my choice to place this illustration in the front cover.

Equation written by Benjamine Betts, in relation to the image, depicting consciousness.

I would like to dedicate this book to my three children.

Who I hope will have a long life ahead of them.

As I have learned by now, the more you live, the more storms you have to navigate.

They may or may not read this book now, but one day, when they are faced with many of the waves that are bombarding the minds of humanity, they will definitely do so.

I might or might not be there. Still, they will be able to feel I am talking to them and presenting them with my views and answers for questions they might have wished to ask me.

They may then realize they were the only reason I wrote this book in the first place.

CONTENTS

NOTE:

This book is neither about religion nor science rather about logic, thinking and philosophy.

A human being's life from his inception till his demise and whatever follows afterwards is a serious business; discussing this over a cup of tea as a philosophical exercise trivializes it and does not represent its real solemnity.

The problem lies with the human brain; it always strives to feel comfortable, and the only way it can do so is by facts (Science), the validity of facts (Logic) and the way they interplay together in the brain (Philosophy).

Thus, philosophy it is on two conditions; first is a reader should approach this endeavor as his own personal future, as one adult human cannot be responsible for the outcome of another, and second is one's aim should never be arguing the validity of a philosophical exercise per se, instead finding the truth.

Can science lead to faith?
I have always thought the answer could be,
Yes, for some people and No for others.
The truth is, I really didn't realise the depth of the question till
I finished going through the logical process you are about to read.
Let me present the reader with my deductive reasoning,
and then later, we can re-address this question again.

Preface

The topics of science and faith have been extensively discussed and presented in the literature.[1][2][3][4] In some contexts, it focuses on scientific revelations of religion and how to enforce the believer's faith using science. In another, mainly in academia[5][6] It revolves around whether science explains all and whether faith has a place in human life. Both sides present logic that could only be described as eloquent. Personally, I felt I could contribute more facts to the argumentation process.

Other views looked at how to reconcile the differences some perceive exists between the two topics, reassuring their readers they can have both in their life.[7] On the other extreme, some academics tried

---

[1] Science and Faith: A New Introduction by John F. Haught Mahwah, NJ: Paulist Press, 2012.

[2] What Your Body Knows About God, Rob Moll. Downers Grove: InterVarsity Press, 2014.

[3] Mapping the Origins Debate, Gerald Rau. Downers Grove: InterVarsity Press, 2012.

[4] The Wonder of the Universe, Karl W. Giberson. Downers Grove: InterVarsity Press, 2012.

[5] God and Stephen Hawking: Whose Design Is It Anyway? by John C. Lennox, 2011. Lion Books.

[6] The Grand Design, by Stephen Hawking, Leonard Mlodinow. Published 2010 by Bantam.

[7] A Leap in Science and a Step of Faith: Seeking God for the Scientifically Curious, by Daniel S. Zachary. Independently published, 2020, ISBN-10: 1654195634

methodically to explain why humans have faith, implying it was an evolution necessity.[8]

The conflict is hypothetical, conceived by insistence of materialists on only accepting evidence without a "shred of a doubt." A concept which is obviously not available to humans on many topics, but intentionally barred from their perception and consciousness. In our endeavor to navigate the labyrinths of science and faith, we will find out using logic and science that this barring is an intentional attribute of human design. Comparable to the answers of an examination, which are not readily available to candidates trying to qualify for a brighter and foresightful future. Rather they are meant to work hard to unearth these mysteries.

We examine an affliction that is adherent to all humans, during their passage through this life and until their definitive exit to a meta-universe alternative. This is "Unfairness," which is either perpetuated by them, or inflicted upon them. Who is causing this unfairness? Why is it associated or allowed within an Intelligent Design? And what is its repercussion on this life and beyond? The answer to these questions might hold some evidence that can be labelled as "beyond reasonable doubt." Which in my opinion is what the human brain needs, to feel comfortable.

Having developed my understanding of this topic over the years, based on my personal upbringing, my reflection of nature, as well as my educational and surgical training back-ground, I have come to my own set of comfortable beliefs. When I fervently felt I wanted to put my thoughts to paper, I read what academia had to say, and I

---

[8] The "God" Part of the Brain: A Scientific Interpretation of Human Spirituality and God. By Matthew Alper. Published by SourceBooks 2008, ISBN-10: 1402214529.

agreed with a lot of what was being articulated, but on many issues, I could not help but detect bias. This led me to believe every human brain must develop its own understanding of science and faith, as this is not a class in a university. One should not be impressed by a view based on its credentials in the academic world, rather you can find deep understanding and logic in the mind of a farmer ploughing in the field who has peace in his mind and balance in his life.

This book aims to look at the relationship between science and faith from a different perspective, which begins from where everybody else has started but takes it further by identifying, comparing and analyzing different aspects of human design using scientific facts then scrutinizing the design attributes from a philosophical and logical point of view. Deducting, what could be the logical reason behind the differences in the design traits within human beings and the human race? A few hypothetical theories will be presented, together with the premises and evidence for their truths.

Using only logical argument and deductive statements, I will try to bridge the gap between science and faith.[9] This will help me scrutinize my own faith by subjecting it to the basic laws of logic, endeavoring to expose basic deficiencies in argumentation a number of humans have fallen into in the past. I realize I am embarking on a very difficult path. Still, I am willing between me and my laptop to argue the matter to the fullest, using basic scientific and recently discovered facts.

The book also aims to communicate and logically challenge atheistic

---

[9] Regis Nicoll writes about the bifurcation of science and faith, stating that it comes down to this: The materialist operates on the belief that "nature is all there is." The word "belief" signifies something that is not scientifically proven.
https://www.allaboutscience.org/science-and-faith.htm

deliberations which agree on a minimal idea of Intelligent Design but are not taking it further in investigating who the Designer is. This includes agnostics, who think why we need to care and theists, who tend to deviate from science and logic. And finally, avid believers of science who put their complete faith in its premises, believing it can explain all phenomena.

A brief discussion of the tools of argument used in the academic fields of thinking, logic and philosophy will be presented with an attempt to explain how humans ended up having such a variable and extreme forms of beliefs on the topic of faith despite having the same type of data input and such precise scientific facts.

The argumentation process throughout this book will be based on two very important observations; the first is the differences in the design attributes of humans when examined at five cross-sectional levels, and the second is the unfairness humans meet in their life on this earth. We will look for the causes behind this unfairness and the consequences that will supervene.

The discussion will also include a philosophical assessment of our life on earth and predict a possible "Meta-Universe life." Finally, looking at an idealistic requirement of faith without choosing one. Then, a reader can feel comfortable with his own beliefs or start investigating unanswered queries. The ultimate aim is to look for the Designer's attributes solely by observing evidence from design. As well as any possible manuals we may find to help navigate our lives.

Finally, after analyzing all the above facts, we will re-visit the relationship between science and faith, trying to expose the constraints posed by both of them and identifying the possible interrelation.

I

The Design

I begin writing this book with the notion I will probably be the only one who will ever read it, yet this doesn't seem to affect my enthusiasm or stamina in putting my thoughts down.

The appraisal of the relationship between science and faith is very complex. They have always been a world apart and seem at a crossroads from each other.[10] [11]The main rationalization for this is the followers of faith, i.e., believers, have not used laws of logic or reason in their narration. As a result, the laws of logic would classify their product as a subjective rather than an objective idea. This, in itself, is not all bad if other methods of verification are built into their system or deposition. However, this does not exist in the majority of doctrines; the supposition is, logically, the faithful narration can only give us an idea about the people behind this idea rather than the idea or the religion itself.

---

[10] In a dissertation about the conflict of science and faith, John Yegge, from Walden University, highlights the ongoing fight, where some public school districts are forbidden from teaching the theory of evolution, others are forbidden to teach creationism or intelligent design, and others are required to teach both. https://scholarworks.waldenu.edu/cgi/viewcontent.cgi?article=1081&context=dissertations

[11] President Barack Obama announced on July 8, 2009, that he would nominate renowned geneticist Francis Collins to be the new director of the National Institutes of Health; a number of scientists publicly questioned whether the nominee's devout religious faith should disqualify him from the position, controversy over his nomination reflects a broader debate within the scientific community between those who believe religion and science each examine legitimate but different realms of knowledge and those who see science as the only true way of understanding the universe. https://www.pewresearch.org/religion/2009/11/05/scientists-and-belief/

In contrast, uncompromising scientists tend to ignore faith when explaining scientific phenomena, and most tend to fall back to the evolution theory to explain our existence. Despite having many gaps and missing links,[12] It seems, for them, relying on measurable elements and ignoring the unmeasurable is a way out, and it closes some doors that have caused controversies and contentions when opened.

On the other hand, religious theologists assert on linking science to faith. Still, the outcome is a feeble connection most of the time, as it is not using scientific language, thus short of convincing, to the extent many edges towards using vague comments that fall short in scientific methodology and logic. Some even use the notion our brain is short in our ability to comprehend faith. Referencing this claim will be akin to favoring one particular faith over another, which we have agreed to avoid.

Humans have decided to agree on science (not in science but its methodology) but disagree on (and in) faith. The human brain has perceived the spectrum of faith ranging from atheism to polytheism to monotheism. In each, you can delve into an endless classification which is still growing to date. Thus, to get out of any quarrel, I have decided to edge on the side of science, focus on the common, and avoid dividing as much as humanly possible.

Commencing with the atheistic point of view, the common ground one can find between them and other religious ideologies is the concept of "Intelligent Design." This theory holds that the universe and all living things can only be explained by an intelligent cause,

---

[12] Please refer to Darwin's Theory of Evolution, discussed in Chapter III.

SCIENCE AND FAITH

not an undirected process such as natural selection.[13] Through the study and analysis of the design components of various existing structures, it can be concluded the product (universe) is not through chance but natural law, intelligent design, or a combination of both; this concept is generally acceptable to many atheists but not all.

Opponents of the Intelligent Design theory claim it opens a back door for theology. The dilemma is, if you get into a logical discussion, you should be ready to examine all facts and possibilities. If you do so on the condition one argument cannot win over another, then this is false logic.

Intelligent Design sounds like the most obvious thing, but how can one comprehend the depth of any particular design, if one cannot understand how the simplest unit in our own body (the human cell) works or functions? How can you comprehend its complexity if you cannot assimilate how it functions?

I believe the only way humans will be able to link the "Intelligent Design" concept to Faith, is if they manage to find a manual or message from its "Designer."

Opponents of the intelligent design concept should stop and look at the verdict published by the world-leading atheist, the late Professor Antony Flew, who was a pioneer of modern atheism. His famous paper, "Theology and Falsification," was first presented at a meeting of the Oxford Socratic Club chaired by C. S. Lewis and became the most widely reprinted philosophical publication of the last five decades. He then did a recourse and published his thoughts in the

---

[13] Intelligent design refers to a scientific research program as well as a community of scientists, philosophers and other scholars who seek evidence of design in nature. https://intelligentdesign.org/whatisid/

book "There Is ~~No~~ A God."[14] in 2007. This presented a detailed account of his decision to revoke his previous beliefs and argue for the existence of God. This is a story of a reasoned thinker, and where his lifelong intellectual pursuit eventually led him to believe in God as a Designer. Suppose atheists don't build on the work and thinking of previous fellow atheists like Antony Flew. In that case, they can probably never find an answer to anything, as our life is unfortunately not long enough to be able to start from scratch again and again.

All other religious groups agree on the topic of a Designer; the difference between them is which one is God, and going on through this path is futile and endless. Rather, we can look at the criteria we expect to find in the "Designer" by evaluating his product. Thus, it seems "Design" is the common ground we should pursue in our quest to link science and faith.

PHYSICAL FORCES

Before discussing design, we need to mention that scientists have identified four forces in our current universe.[15] Some are trying to identify a fifth force,[16] but let us only talk about the agreeable rather

---

[14] There Is No A God: How the World's Most Notorious Atheist Changed His Mind. by Antony Flew (Author), Roy Abraham Varghese. Published October 2007 by HarperOne

[15] In the article "Fusing Astronomy & Physics" from the University of Chicago Library, The universe cannot be understood if only part of it can be seen; the field of astrophysics developed as physicists and astronomers began to collaborate to understand how the fundamental forces of nature have guided the evolution of the universe. https://ecuip.lib.uchicago.edu/multiwavelength-astronomy/astrophysics/index.html

[16] Lead author Jonathan Feng, a professor of physics and astronomy at the University of California, said in a statement. "For decades, we've known of four fundamental forces: gravitation, electromagnetism, and the strong and weak nuclear forces," Feng added. "If confirmed by further experiments, this discovery of a possible fifth force would completely change our understanding of the universe, with consequences for the unification of forces and dark matter." https://www.space.com/33750-fifth-force-of-nature-dark-matter.html

than the controversial. These forces govern everything, including the lilies in the fields up to the explosion of the galaxies or even the inception of new ones.

A brief explanation of these forces is as follows:

i. *Gravitational force:* This represents the apple falling from the tree; this is a relatively a weak force as you can lift yourself on a staircase and hold an apple to eat against such force.

But don't be fooled, as the bigger an object's size, the stronger is its gravitational pull, so one might not be able to walk on a planet larger than Earth.

Imagine if the gravitational force was a little bit stronger; the result would be stars would form so quickly, galaxies will be much smaller and denser and life will not be possible.

ii. *Electromagnetic forces:* Govern electricity and magnetism, which, in summary, governs the positive and negative charges of atoms from which everything is made, including our human design.

iii. *Strong nuclear force:* This force is responsible for holding the protons and neutrons in the nucleus of atoms together. Considering only electromagnetism and gravity, the nucleus components should fly off in different directions.

iv. *Weak Nuclear Force:* These are forces existing between the elementary particles in an atom; for example, they are responsible for radioactive decay, specifically the beta decay neutrino interactions. It has a very short range (less than one femto-meter), and this force is considered weak in nature as compared to the strong nuclear force. However, it is much stronger than the gravitational force (10^{25} times stronger).[17] This weak nuclear force allows Carbon 14 dating as it decays at a regular rate and thus can be measured.

THE CONCEPT OF HUMAN DESIGN

Working for many years with the human body, studying anatomy, physiology and human diseases, one cannot help but stop and marvel at the concept of our body design.

Humans have an inherent tendency to plan and design; they go into a great depth developing expertise in assessing, evaluating and appreciating design. Indeed, we respect and love intelligent design. You only need to listen to an engineer explaining her design concept of a new car or a machine, and you will immediately see and feel the passion in her eyes and speech. A person who has not worked in the medical or paramedical field might find it intriguing or challenging to appreciate the biological ingenuity of the human body but cannot deny the obvious design concept. We need to look at our body as some form of design at both the microscopic and macroscopic levels.

I will explain here why I have chosen to discuss the concept of the design of the human body rather than any other scientific marvel.

Philosopher and mathematician David Berlinski argues: *The universe is set in such a way, things in it are so precisely coordinated, numerical values for certain parameters are so finely adjusted, and if they are perturbed in any way, the universe would collapse or become chaotic, and living systems could not develop. Thus, we cannot conclude this*

---

[17] https://www.vedantu.com/physics/weak-force

could result from an accidental process. He then compares this to laws of nature that don't work with the same principle but can work with any numerical valuation or range.[18] [19]

This refers to about 26 physical constants which are precisely fixed,[20] like the speed of light in a vacuum, the charge of an electron in an atom and many more. They are so precise in value if there is any change in them, either the universe will not exist or life will not be sustainable. Some has termed it the knife edge of design. The epitome of which is in an antigravitational force which allows the universe to expand despite the pull of gravity from black holes, major planets and galaxies.[21]

This highlights a fundamental design principle; the design process has chosen a strict law to govern one aspect of the design and allowed a little more freedom to govern another (laws of nature,[22] like how plants grow). This certainly should draw the attention of knowledge

---

[18] In an interview by Peter Robinson, David Berlinski talks about "Atheism and its Scientific Pretensions," Recorded on April 25, 2011 and published by the Hoover Institute. David Berlinski is an American author who has written books about mathematics and the history of science as well as fiction. He is a senior fellow of the Discovery Institute's Centre for Science and Culture, an organization dedicated to promulgating intelligent design.

https://www.youtube.com/watch?v=FyxUwaq00Rc

[19] This is also dubbed the fine tuning of the universe for life. Fine-tuned universe.

https://en.wikipedia.org/wiki/Fine-tuned_universe

[20] It Takes 26 Fundamental Constants To Give Us Our Universe, But They Still Don't Give Everything. An article by Ethan Siegel in Forbs magazine lists and explaining some of these constants.

https://www.forbes.com/sites/ethansiegel/2015/08/22/it-takes-26-fundamental-constants-to-give-us-our-universe-but-they-still-dont-give-everything/?sh=56607bd24b86

[21] In a video interview, Physicist Leonard Susskind talks about: Is the Universe Fine-Tuned for Life and Mind?

https://www.youtube.com/watch?v=2cT4zZIHR3s

[22] According to the taxonomy of scientific laws, some laws are deterministic, and an example is the laws of Newtonian mechanics; others are Probabilistic, like the law of Mendelian genetics and of social and economic development.

seekers trying to make sense of our existence and purpose in this life or beyond.

The same principle can be identified in human design, which we will try and analyze in the following section.

Philosophically, the human body is a marvelous machine. Still, contrary to our man-made ones, the human is a model grown from a sperm and an ovum. Through a process of biological equations and chemical formulas and not magic, these two structures unite and start a process of unprecedented complexity that has no comparison in this universe. Not only is this design systematic but repeatable millions upon millions of times; thus, it is not a coincidental or a one-off process. The outcome is an ultimate machine that not only feels and perceives data but processes these data very well. This human design can grow, learn, heal, have awareness, and interact with other designs effectively.

If we try and look at this design in a cross-sectional view at different levels, we may learn more about the nature of our design and the intent of our Designer. It will be clear, as we progress up in the design process's scale, the function and the performance increase in complexity. But what concerns us more here is the character of these designs and what we can learn from them about why we are here and where we are heading.

HUMAN DESIGN

Five Cross-Sections of Human Species Design

Human design starts with an *atom* that aggregates with other atoms to form the *human cell*, which collectively develops with other cells into an *organ*. Multiple organs are grouped to form a *human being*, who, with a multiplication process, constitutes the *human race*.

We need to look at the design process at each of these levels, compare philosophically and physically the behavior and characters at each one, and question why they differ.

1. **Atomic level:** We don't need to start at a smaller level than this, as the atom is the smallest building block in the design process as well as the smallest functioning unit in the existence of the universe and makes more sense in the appreciation of the macro design than looking at subatomic particles.

   If we had an electron microscopic eye, we would not have seen our current shape or form but a collection of atomic structures moving at high speed. The atomic level is governed by the universe's accepted four known physical forces.

2. **Human cell:** The multiple units of atoms are clustered in massive numbers to form a human cell. According to an estimate by engineers at Washington University, there are around 10^{14} atoms in a typical human cell.[23]

   Cells are the basic structural unit of all living humans; they have the following general basic structures:

   Cell membrane for defence and absorption. A nucleus which contains the cell's genetic material (DNA) and controls the cell's growth and reproduction.

---

[23] Helmenstine, Anne Marie, Ph.D. "How Many Atoms Are There in a Human Cell?" ThoughtCo, Aug. 28, 2020,

https://www.thoughtco.com/how-many-atoms-in-human-cell-603882#:~:text=According%20to%20an%20estimate%20made%20by%20engineers%20at,the%20number%20of%20atoms%20in%20a%20human%20cell

Ribosomes: Tiny structures where proteins are synthesized.

Rough Endoplasmic Reticulum ER: Studded with ribosomes and is involved in protein synthesis and membrane production.

Smooth ER: Lacks ribosomes and is involved in lipid synthesis, detoxification, and other metabolic processes.

Golgi Apparatus (or Golgi Complex): Processes and packages proteins synthesized in the rough ER.

Lysosomes: Contain digestive enzymes that break down waste materials and cellular debris.

This is a general overview, and the presence and abundance of other components can vary based on the type and function of the cell which allows it to perform a specific function, like secreting hormones or chemicals, transmitting feelings, or tasting food. According to function, there are 13 basic types of cells in the human body,: Stem cells, Red blood cells, White blood cells, Platelets, Nerve cells, Muscle cells, Cartilage cells, Bone cells, Skin cells, Endothelial cells, Epithelial cells, Fat cells and Sex cells. They can have further subclassification up to 200 types.

3. **Organ Level:** This level of design is achieved when cells are clustered together in a specific department that performs a particular function, like detoxification (kidney) or absorbing nutrition (gut). We start recognizing specific organs as a more advanced level.

Just imagine if our human design were a result of an undirected process. We would not have observed this cross-section level of design where a specific organ is responsible for a particular function; instead, we would have seen clusters of different cells doing different functions but spread out and distributed inconsistently throughout, probably with nightmarish consequences for the medical students and the human race.

Grouping biological cells into functioning units (organs) seems crucial in the fairness mechanism, which we will allude to later.

4. **Human Form:** A collection of all the organs together develops into a self-sustainable unit, the human body, which is the centre of our attention.

At this level, the human body has around 30 trillion cells in the most recent scientific estimates, which is 3 with 13 zeros next to it.[24] The human body consists of 78 organs, counting all the teeth and all the bones as one organ each.

Organs within a human form perform different functions of varying degrees of importance, yet the communication mechanism between them is unprecedented. Mainly, it is self-driven and autonomous. We have never invented or seen machines that can perform unattended for 70 years. Not only does the human body coordinate between its organs effortlessly, but it also seems to know which organs are needed at any particular time. Moreover, it realises which ones are more important and need more care and protection than others.

---

[24] Bianconi E, et. Al. An estimation of the number of cells in the human body. Ann Hum Biol. 2013 Nov-Dec;40(6):463-71. https://pubmed.ncbi.nlm.nih.gov/23829164/

5. **Human Race:** The human body cannot exist alone but in clusters and communities; thus, this would be another level of the ultimate design. Discussion of the human design at this level will serve a very important purpose in our argument about fairness later, or rather the lack of it.

Design Comparison

Now, it's time to return to our chosen cross-sectional levels of human design, compare some of the features, and contemplate how they differ at different levels.

I urge the readers to use the minor headings as a guide and to do their research through existing literature about further evidence and details of the vast amount of work that has been done to explore these topics.

I have chosen to look at four characteristics for every cross section of the design processes: *(i) Efficiency, (ii) Reliability and the chance of error, (iii) Fairness and resource distribution, and (iv) Command and control.*

The reason I have chosen these four parameters is: Each is managed differently at different levels, i.e., one would have an excellent, faultless control system, and others would be left without control. The question is: Has this been omitted by chance, error or by intention? We will look at alternate scenarios. Which is the "What-if the situation was the opposite." Hopefully, this will make the answer easier. (If an omission or lack of control was a haphazard or part of the design process).

To make the review less confusing, some points might not be relevant at some levels. I include a diagram of the next section so the reader does not lose track of the flow of the discussion.

Atomic Level:

1. Efficiency
2. Reliability
3. Resource and energy distribution
4. Command and Control

Cellular Level:

1. Efficiency
2. Reliability
3. Resource and energy distribution (will be included with Organ Level)
4. Command and Control (will be included with Human Being level)

Organ Level:

1. Efficiency
2. Reliability
3. Resource and energy distribution (Cellular and Organ)
4. Command and Control

Human Being Level:

1. Efficiency
2. Reliability
3. Resource and energy distribution
4. Command and Control (Cellular and Human Being)

Human Race:

1. Efficiency
2. Reliability
3. Resource and energy distribution
4. Command and Control

Atomic Level

Every atom is composed of a nucleus and one or more electrons bound to the nucleus. The nucleus comprises one or more protons and several neutrons. More than 99.94% of an atom's mass is in the nucleus. Electrons cannot be described simply as solid particles. An analogy might be of a large and often oddly shaped "atmosphere" (the electron) distributed around a relatively tiny planet (the atomic nucleus) and is termed the electron cloud; this is what gives shape to all matter in the universe, but we will come back to this analogy on a later note. The protons and neutrons make up each atom's nucleus are just one femtometer. The electron itself is no more than 1/10,000th the size of a proton or neutron. Does this mean atoms are 99.9% empty, as was taught a long time ago? The answer is no. In reality, the individual electron is spread out over an entire atom like a cloud. There are also many subatomic particles (More than 200 subatomic particles have been detected so far, and most appear to have a corresponding antiparticle). Thus, the space is filled with a wide variety of particles and antiparticles.

1. *Efficiency:* At this level, efficiency is at its maximum possible; the stability of an iron or calcium atom is undoubtfully eternal, and the atomic bonds never wane.

2. *Reliability:* It is so reliable, as it is unheard of, that protons and neutrons of a calcium atom would fail without any outside intervention at an atomic level. The chance of error at an atomic level is un-calculable.

3. *Resource and energy distribution:* Without getting into details of quantum physics, electrons orbit the nucleus for eternity; it can jump from one orbit to another if excited, but no atom needs energy or fuel; its electromagnetic charge within the nucleus seems to keep it going. Thus, atoms seem to function without needing energy forever.

4. *Command and Control:* Atomic and molecular forces are absolutely rigid. Outside intervention using chemical bonds is a form of external intervention through an agent, not an internal failure of the unit itself. In their interaction, they have a rule of chemical bonding, and they exchange electrons in a consistent and repeatable method, albeit dependent on temperature, pressure, or the presence of a catalyst. However, molecules don't attempt to fight back a chemical reaction when prompted to do so, but does it consistently and reliably respond.

Cellular Level

1. *Efficiency:* Human cells are efficient in energy production; they can use different fuel sources: carbohydrates (sugars, starches), proteins, and lipids (fats). But they are energy dependant and need sustenance, and frequently so.

2. *Reliability and chance of error:* At this level, we start seeing errors in the function far in-between; sometimes these are

a result of human or outside influence like diet, abuse or malnutrition, as well as viral infections, and sometimes not, like the aging process, genetic errors or errors in metabolism.

One important physical feature of a human cell is aging and cell death, without which the human race would not have survived. This has received an extensive amount of attention in research funding and work. Although cell death does not happen to all human cells at an equal rate, it usually spares nerve cells. Without cell death, accumulating errors in function will lead to the production of abnormal cells, including cancer cells; thus, if there is an error in the death mechanism built in the human cell, this will result in malignant transformation. In layman's terms, Our body produces cells with errors every day (Cancer Cells). Our defence mechanism hunts and kills them on a daily basis, and if it fails to do so, we develop cancer. That is why cancer is more prevalent in the elderly when the defence mechanism is weaker or when the body's immunity is compromised.

Another form of cell death is autophagy,[25] Where the cells get destructed by the body and used for energy. The miraculous part is, the abnormal, diseased, or aging cells get destructed first, which is another form of efficiency. This is what led to the new fashionable idea of "Intermittent Fasting" as it prompts the body to destroy older cells for fuel, and when feeding restarts, rebuilding brings new cell lines. Thus body rejuvenation takes place.

---

[25] Autophagy: cellular and molecular mechanisms. Glick D, Barth S, Macleod KF. J Pathol. 2010 May;221(1):3-12. doi: 10.1002/path.2697. PMID: 20225336; PMCID: PMC299-0190.

Thus, it is quite evident cell aging and death are predetermined in the design process as the mechanism is carried in the cellular DNA and is not a random process.

3. *Fairness and resource distribution* will be discussed at the organ level as it is more relevant than at the cellular level.

4. *Command and control* are more appropriate for discussion at the higher command level. Neither cells nor organs can command things unless they receive an order from the highest command centre (i.e., human level).

Organ Level

1. *Efficiency:* As noted earlier, similar cells are grouped into organs (some are not, for example, blood cells); this allows them to function in a group which makes control more efficient, like increasing or decreasing fuel delivery for increase or decrease in function. The most striking example is the human kidney as a detoxification factory; if we attempt to replace the human kidney function for only 12 hours, it requires very bulky equipment that consumes way more power and at a massive cost, yet energy metabolism research showed the amount of energy needed by the human kidney is minuscule (10% of the energy expenditure of the whole body) which works as an average of 250 K Calories per day.[26]

2. *Reliability and chance of error:* We see errors in the function far in between, and for a moment, we need to eliminate

---

[26] J Mårtensson, Renal Energy Consumption and Metabolism, Critical Care Nephrology (Third Edition) Chapter 10, Elsevier, 2019, Pages 59-64. ISBN 9780323449427.

abuse or infection by another factor, like bacteria. The breakdown rate at the cellular level is relatively higher than at the organ level; this ensures organ survival even if many cells are damaged or lost. This makes the rate of reliability of organs higher than that of individual cells. It is favouring human survival.

Organs have a reserve ability to function beyond their usual needs. For example, a 20-year-old's heart can pump about ten times the amount of blood needed to keep the body alive. After age 30, an average of 1% of this reserve is lost yearly. The biggest changes in organ reserve occur in the heart, lungs, and kidneys.[27] Reliability at the organ level, although superior compared to human-made machines, is not absolute; a human heart beats three billion times in 80 years of life, the highest among all mammals. Although it is a very impressive performance, however, it eventually fails.

3. *Fairness and resource distribution:* Cells among the same organ tend to be treated fairly, i.e., they receive equal nourishment as all the cells in the same organ. However, when an order comes to sacrifice cells for fuel, preference is given to the old, diseased, or cells with DNA errors. This is a smart choice mechanism which seems to favour organ survival over individual cell survival (relative fairness).

4. *Command and control:* Neuronal and hormonal control have absolute fairness when dealing with cells within a particular organ; however, organs are not treated equally, and the

---

27 Aging changes in organs, tissues, and cells: MedlinePlus Medical Encyclopaedia
https://medlineplus.gov/ency/article/004012.htm

SCIENCE AND FAITH

selective turning on-off in function between them seems to be so advanced and controlled by the brain and neuronal signals, according to the different situations the human body is likely to face. Therefore, we will leave these examples to be discussed at the human level.

Human Level

1. *Efficiency:* The efficiency of the human body in converting chemical potential energy into useful work is known as the mechanical efficiency of the body. The body's mechanical efficiency is calculated as a percentage:

2. $$Efficiency = \frac{Work \ X \ 100}{Chemical \ Potential \ Energy \ Used}$$

Our bodies are not 100 % efficient at converting food energy into mechanical output. They are about 25% efficient, this is better than most cars, which are usually around 20%.[28]

Human-Level overall efficiency seems to be lagging; this will be more evident if we compare the human form to the animal form, as some share a similar circulatory and metabolic mechanisms. We are far behind in power, speed, vision, and hearing but far ahead in speech, intellectual and brain functions.

In a natural selection scenario (i.e., species developing from one another through evolution), one would expect everything to improve as the design concept progresses, but in reality,

---

[28] Efficiency of the Human Body – Body Physics: Motion to Metabolism, Pressbooks.pub. https://openoregon.pressbooks.pub/bodyphysics/chapter/human-metabolism/

humans, in many aspects, are inferior in power and vision to many animals claimed to be in their lineage. One can argue the human species do not need this extra muscle power or better[29] vision to survive, and this could be logically true in today's world, maybe we don't need animal power to survive today, but only a few thousand years ago, it would have made human beings more survivable. Yet, the design process decided to give these features to many mammals but not to humans. Indeed, there is no evidence ancient human beings had the same power as apes or other mammals, and they never had the visual ability of some animals.

In reality, efficiency drops here compared to other levels of design we looked at earlier; as an example, an Iron atom never loses power and retains efficiency forever.

3. *Reliability* is the probability of humans conducting specific tasks with satisfactory performance. Individual human performance or reliability seems to be extremely variable between individuals, depending on many factors, such as age, state of mind, physical health, attitude, emotions, propensity for certain common mistakes, errors, and cognitive biases. There are many tools and studies to measure and quantify reliability in human performance; however, it is important in this section to mention that academic work specifies the following formula:

Human error is contrary to human reliability, and basically, the human error probability (P(HE)) is described as:

---

[29] This refers to the evolutionary concept that needs lead to the creation of characters or features in species, as is termed natural selection.

$$P(HE) = \frac{\text{number of errors}}{\text{Number of error opportunities.}}$$

The relevant point to link reliability with design is; humans are designed to have to work hard to be reliable, and reliability is not automatically bestowed upon them. They are rewarded with tasks and jobs if they do and become reliable. Throughout their dependence on one another, they mainly look for reliability; it will be further discussed at the "Human Race" level.

4. *Fairness and resource distribution at the organ and human levels:* When we look at a large number of cell units living together in an organ or body, it is essential to look at their behaviour and how they interact from a fairness point of view.

First, let us agree on the definition of fairness. It is defined as the *"impartial and just treatment or behaviour without favouritism or discrimination."*[30] Fairness is a measurable element of interaction and, as we will see, can give us clues about how we are made and what the designer intended for us.

Then, we need to look at the types of fairness. Researchers have suggested many types, whether absolute or relative.

The study of fairness has mainly looked at populations and companies and is always related to financial remuneration, attempting to link it to risk-taking and level of responsibility. In our comparison, we will link it to the distribution of

---

[30] https://www.lexico.com/definition/fairness

resources, which is similar to the fairness in society models. In general, one can find three types.[31]

Absolute fairness (In philosophy, is described as Sameness): where every cell or organ gets equal nourishment as its counterpart. This is only seen at the cellular level within a particular organ and not outside its border.

Fairness based on performance (In philosophy, it is described as Deservedness), and this is not seen along the cellular or organ level; indeed, if one organ works extra hard like muscles, it will receive extra nourishment in order to fulfil its duty, but does not receive an added supply of nutrients even if they don't need it, just because it worked harder.

Both absolute and based on performance fairness are very easy to calculate and are simpler on the design scale. Although used in population management, it can lead to catastrophic sequel if applied at the human organ level, probably the extinction of the human race.

Fairness based on need (In philosophy, it is dubbed: As Needed): This is the most difficult and complex type of fairness; the most striking example is the starvation scenario, where the complex command and control mechanism starts diverting nutrients to the most vital organs necessary for human survival, like brain, lung, and heart and simultaneously shuts down the supply to less important organs like skin and muscles. If the human design was a

---

[31] It's Not Fair! But What Is Fairness? Psychology Today. In this article, author A Dobrin, Professor Emeritus of University Studies, Hofstra University, reviews the three ideas of fairness. https://www.psychologytoday.com/us/blog/am-i-right/201205/its-not-fair-what-is-fairness

simple natural occurrence, then we would have seen the first two types of fairness and not the based-on-need type.

There will be a more extensive discussion about human unfairness later and the consequences we could expect from its occurrence.

5. *Command and control:* At the cellular level, the organ level, and human level, command and control is self-driven, predetermined with a set of codes, for example, the *"fight and flight mode,"* which is an automatic physiological reaction to an event perceived as stressful or frightening. The perception of threat activates the sympathetic nervous system and triggers an acute stress response, this prepares the body to fight or flee. These responses are meant to increase the chances of survival in threatening situations. Science tells us the following actions take place:[32]

   - Pupils dilate to take in more light so you can see better.

   - Breathing increases and becomes deeper so you get more oxygen in your lungs for needed work by the muscles, heart and brain.

   - The heart pumps more blood to the brain to think better and faster and sends more supply to the muscles so you can sprint or move quickly if needed.

   - Awareness is sharpened, and senses are heightened.

   - The skin of your hands becomes cooler as blood is directed to more essential organs in your body.

---

[32] https://health.clevelandclinic.org/what-happens-to-your-body-during-the-fight-or-flight - response/

- Pain is suppressed to the extent most people who had accidents report they only felt pain later and not during the immediate trauma.

- Frequently, people report memory loss of the actual accident as our brain attempts to protect us.

- Digestive, reproductive and growth hormone production and tissue repair are all halted.

The command and control of all of the above actions is under strong effective unconscious control.

Human Race

1. *Efficiency:* At the human race level, the concept of efficiency changes; the most accurate description of being efficient as a society, a country or a group is written on the website of *"Human Efficiency Solutions LLC,"* founded by George Weber, describing it as *"Doing things that make sense."* [33]

   We see efficiency dropping significantly at this level, as the human race behaviour is the most wasteful and most destructive to their bodies, environment, each other, as well as the planet as a whole.

   We are even dumping a lot of junk in the universe. [34] Improvement in knowledge doesn't seem to have improved the situation; on the contrary. An example is the amount of

---

[33] https://www.humanefficiency.com A website that discusses efficiency in politics and government. They state that being efficient is the key to becoming successful as an individual, organization, and, ultimately, society as a whole.

[34] According to NASA, There are approximately 23,000 pieces of debris larger than a softball orbiting the Earth, half a million pieces of debris the size of a marble or larger, and approximately 100 million pieces of debris about one millimetre and larger, any of them can pause a serious risk to satellites or space stations.

destruction to the environment we are doing today despite the increase in the level of knowledge.

2. *Reliability:* Human reliability is the probability of humans performing specific tasks satisfactorily.[35] Like efficiency, it is low at the individual human level but improves collectively at the human race.

You can rely on a human being to go to work most but not all the days or on some humans to go every day, but you cannot rely on a single human to go to work every single day of his life or on all humans to go to work on a single day. The point here is, one cannot rely on an individual human being, but collectively, they can provide good reliability. Thus, they are forced to cooperate.

Why does reliability in human behaviour improve when working in a group compared to individually? This could seem like an obvious thing that is self-evident. But stop for a moment and imagine a society where individuals work very well separately and when they work together, they tend to fail or fight, as seen in some real-life situations. This would have probably changed the depiction of human civilization as we know it. But this has not happened, probably not by chance of occurrence but by the purpose of intelligent design.

If you want more convincing, try and look at some mammals, who can only live in small groups and would only tolerate one male in their pack. While a lot of others can, the difference

---

[35] Advanced Methods of Risk Assessment and Management, Edited by Faisal I. Khan, Paul R. Amyotte

Volume 4, Pages 1-440 (2020). Chapter Eight - Human Factors Risk Assessment. https://doi.org/10.1016/bs.mcps.2020.02.005

all lies in the connectivity and type of consciousness that was built within them.

One can claim modern group working skills is what has created that. I certainly agree, but the Vikings never had group working skills training to occupy a big part of Europe. But they did it as a habitual second nature skill. Later, when humans realized these skill's importance, they started teaching them in schools and universities.

We can observe human's ability to coordinate complex tasks amongst themselves very early-on in history, this started with hunting, progressed to earth and space exploration, and unfortunately has not been short of war and destruction. The coordinated working skills we develop and acquire is only possible because of our neuromuscular and cognitive wiring built in each individual human being, without which we would have failed to do so.

As we get rewarded for reliability, it encourages us to cooperate more.

3. *Fairness and resource distribution:* At this cross-section of our study, and on the issue of fairness, all rules are broken and can be described as chaotic. How can one describe fairness and resource distribution in the human race using a logically valid statement? Most people perceive life as unfair. Thus, it's not only fairness which is important but also the "perception of fairness."

Let us start from the beginning. An online paper published in 2014 by Rachel L Kendal,[36] who is a Senior Lecturer at

---

[36] https://theconversation.com/the-human-race-evolved-to-be-fair-for-selfishreasons - 31874

Durham University, states: *when given the opportunity to share sweets equally, young children tend to behave selfishly, but by about eight years of age, most prefer to distribute resources to avoid inequalities. She adds biologists are baffled by such behaviour as the theory of evolution by natural selection predicts individuals should behave in ways to maximise their inclusive fitness.*

We also need to look at the broader issue of fairness at the human race level, rationalizing it as fairness to oneself, fairness to animals and fairness to the environment. It is quite obvious humans have done badly at all these levels. And I will spend a few paragraphs proving my point.

What are the causes of injustice and significant degradation of fairness in our societies:

1. *Perception of fairness:* Although there could be no injustice to a particular human, perception of unfairness can make him/her be convinced otherwise, which sets a cascade of events in motion, i.e., leads to action on the part of the individual, which might or might not be justified, this will propagate in the society into bigger and bigger waves of unfairness.

2. *Equity Theory of Motivation (ETM)* was developed in 1963 by John Stacey Adams.[37] It states if an individual identifies an inequity between themselves and a peer, they will adjust the work they do to make the situation fair in their eyes.

   The theory contains two primary components: inputs and outputs. An input is a contribution one makes to receive a

---

[37] A Guide to Equity Theory of Motivation. By Indeed Editorial Team. October, 2019
https://www.indeed.com/career-advice/career-development/equity-theory-of-motivation

reward, while the output is the compensation an individual receives as a direct result of the input they provide. An employee will evaluate his input against what he receives as output. If they think the output is not fair, they will reduce the effort or contribution somehow to the work environment till they feel self-satisfied about the balance of input vs. output. The difference between the ETM and the perception of unfairness, is that in the ETM, there is calculable evidence of unfairness; an example is more pay or promotion despite putting in the same number of working hours, a scenario we are probably all familiar with.

3. *Relative deprivation theory,*[38] which is often attributed to the American sociologist Robert K. Merton, and it suggests people who feel they are being deprived of something considered essential in their society will organize or join social movements dedicated to obtaining the things from which they feel deprived. This action in the past often led to confrontation, which was sometimes violent; this explains some of the riots and violence society confronts repeatedly.

4. *Deviation from normal human behaviour* or possessing a personality disorder.

5. *The law.* Humans have devised laws to control their societies and to achieve fairness; however, throughout history, many laws were made that instituted unfairness and even discrimination and injustice. One stark example of humans creating rules which led to disasters can be demonstrated

---

[38] All About Relative Deprivation and Deprivation Theory, By Robert Longley. https://www.thoughtco.com/relative-deprivation-theory-4177591

by what is known as the "Cobra Effect." During the British governance of colonial India, Cobra snakes increased in Delhi, the Capital and became a menace. The local government made a decision to put a bounty on the species. Probably, the money was so lucrative thus many people hunted the snake till its numbers started to dwindle. But the locals would not have any of it, and they started breading Cobras in order to get the rewards. The pendulum of fortune swung to the other end, when the local government noticed there were fewer cobras evident in the city, yet they were still paying the bounty to the same degree as before. Logically, they stopped paying and abolished the law. Swinging again to the opposite side, the breeders were faced with a massive number of snakes which served no purpose; thus, they let them loose. The city then had a bigger Cobra problem after the law than before it was issued. This is now known as the "Cobra Effect."

Edwinlt, who is a security researcher at the National University of Singapore, on his blog in 2013,[39] mentions three examples of unjust laws, which I would quote only one here as it is the most common globally and the most striking:

Fiat currency law: A bizarre and unjust law is the notion a government has the right to regulate how an economic transaction between two parties of consented adult citizens should be done. In reality, in every country, the government can declare worthless paper notes.

Governments want to monopolize currency because of control. They

---

[39] List of most common unjust laws in our society. Posted on January 12, 2013 by Edwinlt https://indoexpatriate.wordpress.com/2013/01/12/list-of-most-common-unjust-laws-in-our-society/

want to have the power to decide the amount of money circulating in the market. Governments, through the central banks, are able to print money as much as they want to fund expensive and unpopular activities, such as financing wars on the other side of the world or financing bailouts to big banks who sponsored political campaigns against the will of the people.

Printing money will cause inflation and raise the price of commodities such as food, electricity and oil. These commodities are essential to common and poor people. But this law, like all unjust laws, prioritizes the interest of political and corporate elites to finance their wasteful public policies (war and bailout) over the interest of the common people. If authority can decrease the value of your money until you cannot eat, then there is little, if any, reason to become a law-abiding citizen. End of MR Edwinlt quote.

Fairness is the most important topic around which this book revolves, and it was addressed in the organ and human level and here in the human race level. If I can't rationally prove the world we live in is not fair and trace back the reason of this unfairness, then the logical premise I am basing my argument on is false.

So, contrary to the classical meaning of fairness, which is a fair distribution of resources, let's look at other examples of extreme lack of it and who is causing them.

Unfairness to oneself: Can we be unfair to ourselves? The answer is yes and quite often, to summarise this in a logical phrase would be: We usually make intentional decisions and actions we are certain in their detrimental effect on our physical and mental well-being.

In my opinion, the most striking example is smoking, which I failed to find any benefit from, as well as causing grave harm to the human race as a whole, according to scientific research. Yet, it continues to be a significant section of the economic budget and on the balance sheets of mega-corporations and countries. So, governments have decided not to allow you to kill thy neighbour, but would allow you to sell him cigarettes so he can kill himself. Discussion of the deleterious effects of smoking will come later when we look at the causes of human disease.

Unfairness to the animal kingdom: It is a nice exercise to blame fishermen for over-fishing, even though the fish gets eaten. Or whale slaying even if the meat gets consumed. There might be an argument on both sides of these dilemmas. However, I believe the two most stark obvious unfairness against animals are the slaying of an elephant for a tusk that will be of no real value to humans in any meaningful way and the killing of sharks for their fins for a scientifically false claim of benefit to humans. It's utterly heartbreaking.

Unfairness to the environment: I am sure the reader's mind is ahead of me in providing examples of this kind of unfairness, and no, I don't believe humans using fossil fuel is the worst part, again; I will choose the issue which is likely to affect the whole planet, which in my opinion is "The Great Atlantic Garbage Patch." Here are some of the facts about it :[40]

- It Carries 7 million tons of weight.

- Twice the size of the US state of Texas.

- Up to 9 feet deep.

- In the Great Pacific Ocean Gyre, there is 6 times more

---

[40] A company, "Reef Conservation International," is Located in Placencia, Belize. Have an online blog discussing reef conservation.
https://reefci.com/2018/08/11/what-is-the-great-pacific-garbage-patch/

plastic than plankton, which is the main food for many ocean animals.

- By estimation, 80% of the plastic originates from the land, floating in rivers to the ocean or is blown by the wind into the sea.

- The remaining 20% of the plastic originates from oil platforms and ships.

- According to scientists, it is the largest plastic dump on earth, so plastic patches are larger than waste dumps on land.

- Trash patches are made 80 per cent out of plastic.

- Scientific research from the Scrips Institution of Oceanography in California, U.S. shows 5 to 10% of fish contain small pieces of plastic.

This is an environmental disaster of epic proportions, not because of the above-stated facts alone, but because of the spiral growth in size without any appreciable effort to reverse the calamity. One has to acknowledge the efforts of some humans[41] who are fighting hard to rectify the situation; the predicament is those who are creating it do not even know about its existence.

4. *Command and control At the human race level:* Totally opposite to previous levels, this seem to be under full human control. It will be governed by the consciousness of society as a whole, or what is termed "Collective consciousness" or "Collective intentionality." Using the Stanford Encyclopaedia of Philosophy:[42] *Is the power of minds to be*

---

[41] The Ocean Cleanup is a non-profit organization that is developing and scaling technologies to rid the world's oceans of plastic. https://theoceancleanup.com/

[42] https://plato.stanford.edu/entries/collective-intentionality/

SCIENCE AND FAITH

jointly directed at objects, matters of fact, states of affairs, goals, or values. Collective intentionality comes in various modes, including shared intention, joint attention, shared belief, collective acceptance, and collective emotion.

The modern concept examples include solidarity attitudes, memes, extreme behaviours like group-think and herd behaviour, and collectively shared experiences during collective rituals. Rather than existing as separate individuals, people come together as dynamic groups to share resources and knowledge. It also explains how an entire community comes together to share similar values. This has been termed *"hive mind," "group mind," "mass mind,"* and *"social mind."* [43]

But it's here when problems start. According to Plato[44] (428 BCE – 348 BCE), A city-state with good command and control would end up having conflicts with other city-states, as each would attempt to expand and take from other city-states what they possessed. War is inevitably to come from this.

When effective command and control and collective consciousness are established at a higher level between city-states, modern examples are state alliances we see in Europe, America or the Asian continents, this helps establish peace and security within continents.

Wishful Thinking: If we can imagine the human race can agree on command-and-control principles, as well as a collective

---

[43] https://en.wikipedia.org/wiki/Collective_consciousness

[44] Plato's Command-and-Control Utopia: Plato never answered the question: Who guards against the guardians? 2016. An online article by R. Ebeling. A Professor of Ethics and Free Enterprise Leadership at The Citadel, South Carolina. https://fee.org/articles/platos-command-and-control-utopia

consciousness, among continents, then we can see more peace and stability on planet Earth, but obviously, the human race has failed at this level so far.

Before we end the design section, we need to include an essential non-physical part of human design, "Human Psychology." Psychology has a direct effect on how humans behave and some features of it are relevant to our discussion, because of how they help formulate our belief system.

PSYCHOLOGICAL ASPECT OF HUMAN DESIGN:

The design process not only bestowed on the human a perfectly balanced physical body but also installed some behavioral qualities to allow this body to function efficiently. Although many of these characteristics are acquired by learning as we go along, some seem to be born with us and are called essential or inborn traits.

Two important concepts are "Free Will" and "Locus of Control." Both are unique individual characteristics only humans possess (or at least is what is thought now) and should be clear in the reader's mind before we try and link their influence on science and faith.

The third is a phenomenon which seems to characterize all living creatures, "Consciousness," The latter has baffled humans ever since their existence and, for many, bears the key for solving some of the riddles that has haunted philosophers, theists and atheists alike.

Free Will

Free Will in humans is defined in the Encyclopedia Britannica as: *"The power or capacity to choose among alternatives or to act in certain situations independently of natural, social, or divine restraints.*"[45]

This, however, couldn't be any further away from the truth from the philosophical point of view. Although humans believe they are free, this is not entirely true. In philosophy, one can find arguments for and against the existence of free will. I will present the view of Professor Seth Lloyd,[46] [47] which is closest to my belief and the current model of argument.

Humans are influenced by many factors that affect their decisions, and these factors fall under the categories of fear, pleasure, needs, social or religious constraints, past experience and many more. Thus, they are not entirely free to do what they want.

A human decision to eat a piece of chocolate could be decided by their fear of gaining weight, suffering from diabetes or carving the pleasure and enjoying the taste. Thus, we are not free to do anything we want, but our brain enters a computational race where it inhibits one factor in favor of another, leading to a decision. In this particular example, it could be not eating the chocolate. This brain computation takes part way before the decision is made; most of it is subconscious, and our choices are usually repetitive, thus predictable.

The subject is extremely complex. However, I conclude here humans are relatively free and not absolutely free. And have developed the "Illusion of Freewill," but in reality, humans can "Act with a Will."

To prove this, just remember, humans have no control over most

---

[45] https://www.britannica.com/topic/free-will

[46] Lloyd, S. (2012). A Turing test for free will. Phil. Trans. R. Soc. A, 370, 3597-3610. DOI: 10.1098/rsta.2011.0331.

[47] A Turing Test for Free Will by Seth Lloyd. Nov 2016 Seth Lloyd at FQXi's 5th International Conference. https://www.youtube.com/watch?v=5wyJlUUEpSE

of what happens to their bodies or life. They are born in their social class without their choice; they didn't even choose their names, parents or race. As a consequence, their probability of getting a good education, earning money or the most serious feared event in their life, which is when they will die. Thus, the window of relative free will exists but is very limited.

In reality, all humans are moving across their own paths in life, in which they are faced with repeated crossings; at each of these crossings, there is a right or left choice, and each one of them make decisions about which path to choose, according to the factors we discussed earlier. Usually, most of the decisions they make are predictable, and this proves the point.

A big part of our brain computations, which leads to "Acting with a Will," depends on our Locus of Control, which we will discuss next.

Locus of Control

Julian B. Rotter developed the "locus of control" concept in 1954,[48] which plays an important role in personality studies today. This characterized people into two groups: Intrinsic and Extrinsic.

A person's "locus" is classified as internal (a belief one can control one's own life) or external (a belief life is controlled by outside factors which the person cannot influence, or that chance or fate controls their lives).

Locus of control is primarily determined by genetics. However, scientists believe early caregivers of a child can contribute to its development,[49] most believe it is very difficult to change by training.

[48] https://en.wikipedia.org/wiki/Locus_of_control

[49] https://www.psychologytoday.com/us/basics/locus-control

However, it can change naturally over time. Extremely disrupting events in life and unfortunate incidents increase externality.

This topic, at first glance, would seem totally irrelevant to our discussion of science and faith, however when you realize that part of the computation process in your brain which helps you make a decision to exercise your "Will" can be influenced by your locus of control, and in turn, your ability to believe in science or faith.

Let's take some of the examples which have been demonstrated by research. A person who has an Internal locus of control[50] is an individual with the belief if they work hard, they will succeed. Internals tend to lead happier, healthier and more fulfilling lives.

An external locus of control in a person will make them attribute promotion or success to external or environmental factors, such as luck, fate, timing, other people or some kind of divine intervention.

Internals have been found to outperform externals, both in intentional and incidental learning. Internals can retrieve pertinent information and adequate cues that subsequently facilitate their performance and judgement.[51] [52]

Very few people are singularly internal or external. The "Rotter Internal-External Locus of Control Scale"[53] places people on a

---

[50] C Drew, PhD, in July 2022, wrote an online article titled 21 Internal Locus Of Control Examples, explaining the concept and practical examples of human behaviour that would be classified as Internality. He then gives examples of famous people who are classified as internalists. https://helpfulprofessor.com/internal-locus-of-control-examples/

[51] Intrinsic locus of control. Journal of Business and Management. Volume 19, Issue 7. Ver. VII. (July 2017), PP 29-35.

[52] Locus of control, psychological empowerment and intrinsic motivation relation to performance. May 2015, Journal of Managerial Psychology 30(4).

[53] The full 20 question scale, with an explanation of how to obtain the results and to classify an individual human's locus of control, is available on this site. https://studylib.net/doc/8091759/locus-of-control

caliber, on one end the extreme external and at the other the extreme internal. This scale is used as a test for individuals to show which way they lean and if they have a sense of balance between the two extremes.

Although externalists tend to attribute their performance to external factors, specifically here, I would like to focus on "Devine Intervention," as this is mentioned consistently in the literature.[54]

A study in 2020 by Caven et al. and published by Frontiers of Psychology,[55] Examined around 20,000 men and women in the UK for the association between self-reported religious beliefs, attitudes and behavior and locus of control as defined by Rotter 1966. They found people who believed in a divine power were more likely to be internally oriented compared with those who were unsure (i.e., agnostic). And agnostics, in turn, were more likely to be internals than non-believers (atheists). The differences between the believers and agnostics were much greater than between agnostics and atheists..

They then concluded there is a strong association between greater religious belief and a greater internality in both men and women in this population; they also reported on longitudinal data, which showed a non-believer is more likely to become a believer if he or she is internally oriented.

This last major study, in terms of sample size, is totally opposite

---

[54] Locus of Control, Religious Adherence, and God: Who's in Control? J. Haase, Th.D. April 2018.
https://www.researchgate.net/publication/324138580_Locus_of_Control_Religious_Adherence_and_God_Who's_in_Control

[55] The relationship between locus of control, religious behaviour and beliefs in a British population. Caven et al. Bristol Population Health Science Institute, Bristol Medical School. Article number1462. Journal Frontiers in Psychology. Volume11, 25 Jun 2020.

the common belief that religious people will attribute the events in life to acts of God; thus, they are most likely externalists. I think this demonstrates that blaming divine power for bad events and problems in life by Externals, is an excuse rather than a belief.

Consciousness

There seems to be a big problem in finding an acceptable universal definition of consciousness, and where does it ascend from? A simplistic definition is:[56] *The individual awareness of unique thoughts, memories, feelings, sensations, and environments.* Essentially, it is the awareness of oneself and the world around you.

The dilemma is, how does 1.3 Kilograms of flesh (3 pounds of weight), which we call the "Brain," transfer input signals into feelings and emotions, which then makes you realize who you are, gives you an individual personality, is beyond any human's ability to explain or understand, but try they have....

One way to understand consciousness is to examine our state when we lose it (either during sleep, dreaming and in patients who are unconscious).

Theories of Consciousness: Neuroscience believes it arises from a connection of neurons in the brainstem called the "Reticular formation."[57] Its function involves a broad range of autonomic, sensory, motor, behavioral, cognitive, and mood-based responses. It works with other central nervous system regions to allow complex tasks such as the regulation of our state of consciousness, emotion

---

[56] https://www.verywellmind.com/what-is-consciousness-2795922#toc-theories-of-consciousness

[57] https://www.youtube.com/watch?v=UwXQSNcytTY

processing, and visual coordination. Approximately 100,000,000 impulses are received in the reticular formation every single second.

The Classical Computational Theory of the Mind:[58] According to this theory, cognition is a Turing-style computation over mental symbols.[59] Thus, it deducts that cognition can be realized (i.e., recorded and interpreted) not only in the brain but also using silicon chips. Artificial Intelligence scientists who believe in this explanation tend to think one day, they can create machines with consciousness.

On the other hand, Sir Roger Penrose, a British mathematical physicist awarded the 2020 Nobel Prize in Physic, believes consciousness is an un-computational process. He came up with the Orchestrated Objective Reduction theory of the mind together with Stuart Hammeroff, who is an anesthesiologist at the University of Arizona, USA. He claims the answer to consciousness may lie in a deeper knowledge of quantum mechanics.

The non-computational idea is shared and defended elegantly by Riccardo Manzotti from the Institute of Consumption, Communication and Behavior, IULM University, Milano, Italy.[60]

He writes: *"It may be misleading to take information to be similar to phenomena like digestion."* He explains information is computational,

---

[58] The theory of information integration suggests consciousness is the expression of a property of information. Tononi, G. An information integration theory of consciousness, BMC Neurosci.5, 1-22. 2004.

[59] Michael Rescorla. The computational theory of mind. In E. N. Zalta, editor, The Stanford Encyclopaedia of Philosophy. 2015 edition.

[60] The computational stance is unfit for consciousness, International Journal of Machine Consciousness. International Journal of Machine Consciousness. Vol. 4, No. 2. 2012. P 401-420 R Manzotti.

https://www.researchgate.net/publication/263903602_The_computational_stance_is_unfit_for_consciousness but the impulses the brain receives are not information; the brain gets signals, as in reality, information does not exist; it is only in the form of zeros and ones. He concludes: *Thus, computational theory cannot be applied to consciousness.* This group of scientists believes, with the current technology, science cannot create a conscious machine.[61]

I will leave it to the reader to decide which way he believes, but I personally find the non-computational theory of consciousness more plausible. Reducing the human brain design to a Turing style machine is an oversimplification of the process; furthermore, much of the signal processing taking place in our brains is done unconsciously,[62] Thus, signal processing and consciousness could be two separate processes.

Suppose one thinks the human body's design is intelligent. In that case, installing consciousness in this design is the epitome of ingenuity as it has transformed human beings into unique individuals and not just replicas of machines coming off the labor ward in hospitals. We can understand logically how humans function, yet to date, we have no clue how consciousness is realized or produced.

It is believed animals have a consciousness, but Aristotle thought only humans have a higher form. It shows in the form of the ability to have conceptual thought and language.

An American philosopher, Thomas Nagel, wrote a paper, *"What Is It Like to Be a Bat?"* It was first published in "The Philosophical

---

[61] Signorelli CM. Can Computers Become Conscious and Overcome Humans? Front Robot AI. 2018 Oct 26;5:121.
https://www.ncbi.nlm.nih.gov/pmc/articles/PMC7805878/
[62] https://philarchive.org/archive/CRAUBA

Review" in October 1974.[63] The paper describes several difficulties posed about consciousness. What concerns us here is what it means to be a particular, conscious thing like, for example, a bat. He claims we can try to understand some logic of how to be a bat, but only if we have the bat consciousness can we truly realize it. Nagel claims even if humans were able to metamorphose gradually into bats, their brains would not have been wired as a bat; therefore, they would only be able to experience the life and behaviors of a bat rather than the mindset.[64]

Try a closer exercise by putting yourself in another human being's situation. You can draw similarities to this particular person only if you have experienced the same scenarios as he did, yet your response will be different; this proves that the mindset, i.e., the consciousness of each human, is unique. But can you imagine his consciousness or mindset if you have never experienced the same encounters?

When one faces a new experience, he has only heard about previously (an example is the loss of a close relative), his perception of this experience totally changes upon experiencing it himself, rather than hearing stories about it. This is our unique consciousness that is different in every human compared to another, which is formed by our thoughts, past experiences, ideas and beliefs.

Consciousness and Faith: How do consciousness and faith interplay in our brain?

As we are exposed to events or information about a multitude of things, let's say an example related to our discussion here, topics of faith. This triggers short-term memory, which involves relatively

---

[63] The Philosophical Review LXXXIII, 4 (October 1974): 435-50.

[64] From the website of Dr Peter Sjöstedt-Hughes, Philosopher of Mind and Metaphysics. Based on Thomas Nagel paper – What Is It Like to be a Bat?

http://www.philosopher.eu/others-writings/nagel-what-is-it-like-to-be-a-bat/

quick and simple chemical changes to the brain's synapses (narrow gaps between nerve cells).

To build long-term memory, neurons (nerve cells) manufacture new proteins in the brain synapses. Using our consciousness, we subject this information to analytical judgment in our "Occurrent Consciousness." We then realize this information and try to verify its correctness; once we are convinced, it starts to build in the form of belief, which consolidates into faith. We need to constantly be accessing our faith by recalling it into our Access Consciousness so it guides our life decisions and actions.

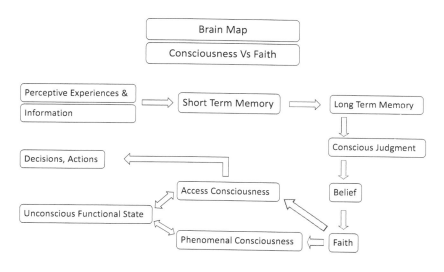

Figure 1: Brain Flow Chart

Access-consciousness: is consciousness availble for use in reasoning and rationally guiding speech and action.

Phenomenal consciousness is experience; the phenomenally conscious aspect of a state is what it is like to be in that state. Sensory experiences such as hearing, smelling, tasting, and having pain.[65]

---

[65] On a confusion about a function of consciousness. Journal of Behavioural and Brain Sciences. Volume 18, Issue 2. June 1995. pp. 227-247.

The brain Flow Chart in Figure 1 is a simplistic way of explaining current knowledge. It is important to note, in every direction of flow in this chart, we are likely to acquire errors, like gathering corrupt data, suffering errors of judgment, poor recording of memory, and degradation of long-term memory, as well as being affected with cognitive bias. This can explain many phenomena we see in real life, like fundamentalism, racism, intolerance as well as extremism. These are all beliefs formed with errors building up along these pathways, but their owners seem to be absolutely convinced of their validity.

When a brain is convinced it has reached the correct destination for a believe or a faith, it starts to protect it. This is guarded by a brain mechanism that initiates fear to defend this faith. I.e., the brain begins to induce fear when it feels its belief/faith is confronted.[66] This explains why it is challenging to correct faith errors when they build up in a human brain.

Research indicates that the posterior Medial Frontal Cortex (pMFC) section of the brain functions as a "neural alarm complex" involved in registering threats. Holbrook et al.,[67] investigated whether pMFC similarly mediates ideological (belief) threat responses, and they found out that downregulating pMFC via transcranial magnetic stimulation (TMS) caused less avowed religious belief.

If we translate this medical jargon, it means; researchers identified a section of the brain (pMFC) that supports belief in general,

---

[66] Neural correlates of maintaining one's political beliefs in the face of counterevidence. Sci Rep 6, 39589, 2016. Kaplan, J., Gimbel, S. & Harris, S.
https://www.nature.com/articles/srep39589

[67] Posterior medial frontal cortex and threat-enhanced religious belief: a replication and extension
C. Holbrook et al. Social Cognitive and Affective Neuroscience, 2020, 1350–1356.
https://www.ncbi.nlm.nih.gov/pmc/articles/PMC7759203/pdf/nsaa153.pdf

including religious types. When they inhibited this section of the brain by targeting it with a focused magnetic beam (TMS), they inhibited belief. I hope the reader does not think I am implying people who don't have a belief have a problem with this particular section of the brain. "As not believing in itself is a kind of belief."

To appreciate the power of belief on the human brain, one needs to look at a familiar phenomenon most people have come across called the "Placebo Effect." This is when a non-active drug is given to a patient to treat a real disease, and the patient gets actual improvement. It works when your brain convinces your body that a fake medicine is a real thing; placebos work on symptoms modulated by the brain, like the perception of pain,[68] but they will not remove a tumor or heal a fracture. The reader should not confuse this with positive thinking.

R Ornstein reported in his book "The Healing Brain."[69] A case of a woman suffering from severe nausea and vomiting. Objective measurements of her gastric contractions indicated a disrupted pattern matching the condition she complained of. They then told her about a new, magical, extremely potent drug that would, the doctors proclaimed, undoubtedly cure her nausea. Within a few minutes, her nausea disappeared. The same gastric tests then revealed a normal pattern when, in reality, she had been given a syrup of "ipecac," which is a substance usually used to induce vomiting.

Skepticisms vs Belief: Skepticism is defined as *"The attitude of doubting knowledge claims."* Philosophically, a sceptic is: One

---

[68] https://www.health.harvard.edu/mental-health/the-power-of-the-placebo-effect

[69] The Healing Brain: Breakthrough Discoveries About How the Brain Keeps Us Healthy, February 1999. by R. Ornstein, D. Sobel. Malor Books; Illustrated edition. ISBN-10: 1883536170.

who, like Pyrrho and his followers in Greek antiquity, doubts the possibility of real knowledge of any kind; one who holds that there are no adequate grounds for certainty as to the truth of any proposition whatever. It is surprisingly easy to generate doubt about human knowledge, even knowledge about the kind of things we consider easily known.

Our brains are wired to use skepticism to question the validity of information and help us discard false or what we perceive as wrong beliefs. Practically, we exercise this process of uncertainty to various degrees, i.e., on one end of the scale, some of us easily fall into believes of supernatural mysticism, and on the other end, some of us tend to exercise skepticism to its extreme, which means no matter how much evidence is presented to them, they are still hardcore non-believers.

Skepticism is an integral part of any believe system, as Michael Shermer argues:[70] *The brain is a belief engine. Using sensory data that flow in through the senses, the brain naturally looks for and finds patterns and then infuses those patterns with meaning, forming beliefs. Once beliefs are formed, our brains subconsciously seek out confirmatory evidence in support of those beliefs, accelerating the process of reinforcing them.*

We can clearly identify three patterns of human belief systems:

The first is where the majority of believers fall into, when beliefs come first, and explanations are pursued afterwards.

---

[70] The Believing Brain: From Spiritual Faiths to Political Convictions – How We Construct Beliefs and Reinforce Them as Truths. Page 4. By Michael Shermer, Publisher: Hachette UK, 2012. ISBN: 178033530X.

https://books.google.ae/books?redir_esc=y&id=a1ueBAAAQBAJ&q=

The second is the balanced believers, who deciphers data by utilizing titrated skepticism and look for confirmation and then believe; we can then infer that this is what the human designer intended from installing skepticism in our brain wiring and consciousness.

The third is getting more popular among non-believers by consciously or subconsciously adapting the philosophical sense of skepticism, where one cannot really confirm the validity of any given information.

My personal view on consciousness: Like many of the references I looked up and papers I read to get a scientific explanation of consciousness. I finally say its conception is unknown. But I must admit I find myself having a deeper understanding of what it is rather than how it works, probably similar to a car or an airplane; most people don't understand how they work. Still, they can be comfortable describing what they are.

Steven Smith,[71] published a book in 2002 titled: *"The Inner Light Theory of Consciousness."* Not only did he attempt to explain consciousness with his theory, but he also tackled the "Mind Body" dilemma,[72] which was initiated by Descartes in the 17th century and has never been settled to date. He favors the computational theory of the mind and tackles the problem at hand from a scientific point of view and not a philosophical one. He explains that our

---

[71] The Inner Light Theory of Consciousness. Steven W. Smith, Ph.D. President of "Tek84 Engineering Group" in California, USA. California Technical Pub. Feb. 2002. ISBN-10: 0966017617.

[72] The Mind-Body Problem and the History of Dualism. The mind-body problem is the problem: What is the relationship between mind and body? Or alternatively, what is the relationship between mental properties and physical properties? Stanford Encyclopaedia of Philosophy.

https://plato.stanford.edu/entries/dualism/#MinBod

brain experiences events and limits the awareness of the process of how we can feel these events but allows them to be consciously felt. An example is one can never be aware of the process which led him to feel hungry, and he only feels the end result, which is hunger.

The DNA is our code, which is translated into organs and functions. Thus, the commonly used phrase that we are wired to do certain things is metaphorically correct. If one day you were brave enough to tolerate any kind of pain, it would be foolish to assume you are so clever; without the wiring built in you, little pain could have seriously disturbed your body systems or could have even killed you.

I believe our consciousness is wired to think, analyze, judge and believe. But we are not wired to believe in anything specific, so you can choose any type of belief you want, even if this is a no-belief (a no-belief is a belief). Thus, we are supposed to fill in the blanks in our life, like believing in family or self, values or immorality, religion or atheism, democracy or theocracy.

Consciousness Limitations: Does the human consciousness have limitations or boundaries? There is some logical evidence to their existence:

- Try and feel your heart beat in a resting situation or the activity of your liver, kidney, or any other function you are sure is taking place in your body at any particular time. You will fail in becoming consciously aware of these functions because your nerves are not connected to do so. Few functions come to our consciousness and we can realise them; many others we do unconsciously, and we are not capable of realising them in our "Access Consciousness."

- Spencer M. Robinson from the Centre for Applied Social Neuroscience[73] (CASN) explains the reason behind the above phenomenon: *The billions of sensory inputs that are constantly being processed by the brain are so vast and complex that they would swamp and totally overwhelm the conscious mind. If such micro-substratum activities were conscious, all concentration would be totally absorbed in analysing how each micro-stimulus is processed and internalized, and we would starve to death dwelling on the sensations and impressions of being hungry rather than forming a broad perception of the world that would enable our interaction with that world to obtain nourishment.*

- The book titled "The Inner Light Theory of Consciousness," which we referred to earlier, attempts to explain consciousness as a form of a limitation or boundary rather than awareness. Steven W. Smith writes: *Our brain has limited ability to perceive its own operation. We do not know how we recognize a face, experience pain, or develop a thought, only that we can do these things.*

This opens the abductive logic: if one boundary of consciousness exists, then there could be more boundaries. Perhaps the way to identify the boundaries of Consciousness is to look for things and issues in the life and history of humanity that are non-physical (i.e., not bound by the 4 physical forces, as if they are, then we expect science to be able to explain them one day) and have never been amenable to our perception and consciousness.

---

[73] The Centre for Applied Social Neuroscience (CASN) is a laboratory for research and development in mental health and education in pursuit of the goals of the new field of Applied Social Neuroscience based in Fukui, Japan. Robinson, Spencer M. 2017. The Limits of Consciousness. 10.13140/RG.2.2.18516.71043.

The limitation of consciousness we have scientifically verified is a very significant step in our attempt to resolve conflicts between science and faith, as some scientists have alluded to the possibility they can one day come up with an answer to everything, and theologians tried to come up with explanations which were not satisfactory to many. The dilemma is the simple human brain followed the theologians for a few hundred years and now is trying its luck with science for a few hundred more years. What we should all grasp is, anything the design process has curtained away from our consciousness, we will never be able to feel or realize.

However, telling humans they should not discuss or contemplate is unbefitting, as the thinking process and understanding are what will help us identify the boundaries of our consciousness.

Consciousness limitation phenomena can also explain the difference between the human higher order and lower order described in animals by ancient philosophers. Besides the smaller size of their brain and fewer neuronal connections, there is also a limitation in the software that curtains their realization of the world around them to a specific purpose.

For example, the consciousness of a pigeon with a tiny brain size will not allow it to comprehend the logic of this book, yet it has a superb realization system to navigate the magnetic north; their consciousness has allowed them some realization but limited so many others. This demonstrates a purpose-design build rather than a sequential upward build from amoeba to humans.

The Heart Brain Duality of Consciousness: In the process of understanding consciousness, I have stumbled upon another divergence between science and faith, which is the heart vs. brain role in its realization. On first note, I didn't want to sidetrack into a less relevant topic; till I realized that pro-science atheists are using multiple old theological claims of heart consciousness as a premise for superiority in the argumentation process.

Thus, we need to dig in deeper for the basis of the scientific evidence outside any discord, avoiding any theological citations, as, by default, it would be classified as a claim.

Let us pose a medical philosophy question: Do we see with our eyes or brain? Scientists will tell you the brain, but when you analyze the narration, they claim the brain is what processes, interprets, and gives awareness to the visual information. The interpretation of the vision and its relation to memory and previous experience happen in the brain, as our eyes register the image upside down and send a signal to the visual cortex of the brain, which then processes the images into the correct orientation then allows us to realize it. Further, the composite image that you see now is not really localized in any one part of the brain. As scientists have identified five centers in the brain, where shape, motion and color are processed in different locations. Thus, seeing with the brain is a metaphorical description. However, we still attribute the function and the conscious realization of vision to the eye.

Do we feel hot and cold with our skin or brain? We feel the temperature in our skin, but if the nerve connection with the brain is lost, you cannot feel anymore; thus, signal processing takes place in the brain, but we consciously realize the feeling of hot, cold and pain in the skin.

Do we feel thirst in the mouth or brain? The answer will be the same, the brain is what processes the signals from various organs then makes us consciously realize thirst in the mouth. The same connection can be made between Hunger and Stomach.

All of our senses are processed in the brain but are consciously realized somewhere else, yet we ascribe these functions to the organ realizing them and not where the processing takes place.

This brings us to our topic of total consciousness; neuroscience has informed us that multiple centers in the brain are important to process our consciousness state. This then leads to the obvious uncertainty: Where do we realize our consciousness?

The answer to this question to date is not known. Claiming it is realized in the brain would be erroneous; as nobody has ever been able to prove that, we can only be sure that processing of the signals necessary for consciousness is processed there; besides we don't really feel our brain; surgeons have operated on the brain of a conscious patient without anesthesia, as brains don't have feelings. Thus, you can't realize happiness or sadness in the brain, despite the fact that the processing of these feelings takes place there, as proved by functional magnetic resonance imaging. While direct stimulation of the brain's reward pathways can induce pleasure, true happiness or a sustained sense of well-being is a more complex construct and is part of the conscious realization.

Can the Heart/ Gut be the center of consciousness realization? Most of us have had this gut feeling or said our heart feels good or bad about a particular issue in life, but this is not science. Let us look at some basic facts:

- The heart is one of three organs that can function without influence from the brain, the others being the intestines and the liver, i.e., if the heart stops receiving any impulses from the brain, it will not stop beating. On the other hand, if the heart stops sending blood to the brain, the latter dies in 4 minutes.

- The Sympathetic chain and the Vagus nerve both connect the heart to the brain,[74] The vagus nerve has more afferent fibres than efferent, which means the heart sends far more signals to the brain than it receives.

- The heart possesses a large collection of neurons, which were shown to have the ability to grow and retain memory; scientists call it "the heart's little brain."[75] The same phenomenon is present in the gut, except that it is much larger. Andrew Armour from the University of Montreal, Canada, first introduced the concept of a functional heart brain in 1991. He later published a book in 2003, "Neuroradiology: Anatomical and Functional Principles."[76] Where he describes the anatomy and function of the cardiac nervous system, as well as the complex neuronal processing and memory capabilities of the intrinsic cardiac neural connections, indicating the heart's brain can process information and make decisions about its control independent of the central nervous system.

---

[74] https://www.kenhub.com/en/library/anatomy/innervation-of-the-heart

[75] The Heart Consciousness – a Neurological Perspective. Jan. 2015.
https://wakeup-world.com/2015/01/22/the-heart-consciousness-a-neurological-perspective/

[76] Neuro-cardiology--Anatomical and Functional Principles, by A. Armour, Published in the USA 2003 by Institute of HeartMath14700 West Park Ave., Boulder Creek, California 95006831-338-8500.
https://www.academia.edu/37169649/Neurocardiology_Anatomical_and_Functional_Principles

In a paper by M. Liester, "Personality changes following heart transplantation: The role of cellular memory."[77] He discusses the role of cellular memory as well as intracardiac neurological memory in transferring memory from the donor person to the recipient patient through the heart. He reports on accounts of recipients acquiring the personality characteristics of their donor, including changes in preferences, alterations in emotions, temperament, modifications of identity, and retaining memories from the donor's life.

In summary, science has demonstrated that the heart has a little brain with an extensive number of interconnecting neurons and ganglions that process signals in the same way the brain does; this seems to go on side by side with the signal processing in the cardiac centres of the brain.

- If consciousness processing takes place in the brain and the heart has exact similar neuronal connection as the brain, it goes without saying that some of the processing of our consciousness can take place in the heart as well, and referring to the transfer of some memory and traits to a heart receipient confirms this hypothesis.

- "Heart Rate Variability (HRV),"[78] a phenomenon identified which shows the direct confluence of our consciousness status with the heart. It demonstrates how the heart coherently responds to our state of mind, which could help

[77] Personality changes following heart transplantation: The role of cellular memory. Liester MB. Med Hypotheses. 2020 Feb;135:109468. PMID: 31739081.

[78] Heart rate variability, Wikipedia. https://en.wikipedia.org/wiki/Heart_rate_variability

in the realization of our consciousness status. It is not known if other organs in the body, like the gut, would have a similar role. Let us look at some details of what does HVR represent:

HRV: is the physiological phenomenon of variation in the time interval between heartbeats. It is under the regulation of the parasympathetic and the sympathetic nervous systems. Variation in our psychological status represents the balance of these two influences (sympathetic and parasympathetic). High HRV represents proper emotion regulation, improved attention and better decision-making; this is associated with increased parasympathetic activity, while low HRV reflects poor emotional regulation, poor attention and bad decision-making and is attributed to increased sympathetic activity. [79] [80] [81] [82]

H. Yang in 2017 published a pilot study.[83] Investigating the differences in HRV indices between two opposite emotion states, happiness and sadness and, it was concluded that HRV indices showed significant differences between happiness and sadness emotion states.

---

[79] Heart Rate Variability as an Index of Regulated Emotional Responding. Review of General Psychology. Appelhans BM, Luecken LJ. Sept 2006. 10 (3): 229–240. doi:10.1037/1089-2680.10.3.229. ISSN 1089-2680.

[80] Claude Bernard and the heart-brain connection: further elaboration of a model of neurovisceral integration. Neuroscience and Biobehavioral Reviews. Thayer JF, Lane RD Feb 2009. 33 (2): 81–88.

[81] Heart Rate Variability and Cognitive Function: A Systematic Review. Frontiers in Neuroscience. Forte G, Favieri F, Casagrande M. July 2019. 13: 710.

[82] Anxiety, attention, and decision making: The moderating role of heart rate variability. International Journal of Psychophysiology. Ramírez E, Ortega AR, Reyes Del Paso GA Dec 2015. 98 (3 Pt 1): 490–496.

[83] S. Yang, L., et al. Differences of Heart Rate Variability Between Happiness and Sadness Emotion States: A Pilot Study. J. Med. Biol. Eng. 37, 527–539 (2017).

Conclusion about consciousness:

Consciousness, in my perspective, is an individually built-in screen through which we observe the world around us. The image you are seeing now of this book is not what your eyes are looking at, as your eyes are recording this page upside down and right to left, but your brain, for an unknown reason, knew to process this image to the correct orientation and projects it to your access consciousness screen. And this is what you are actually observing. This screen has not been identified in any one center in the brain, and it does not shut down when you sleep but keeps displaying images from stored memory (dreams) with no input from your eyes.

Imagine two car models, the first is the usual with a glass windscreen, where the driver is watching directly what goes on in front of the vehicle; the second is a futuristic model, where the windscreen does not exist, and the front is all metal and is part of the car body. Here, there are two super 3D high-definition cameras placed in front of the car; these relay stereographic images, which are displayed on a screen on the inside of the car, where the windscreen is meant to be. The driver, with time, will get so used to this and will adamantly believe he can see everything directly without realizing the difference between the two car models. Our Human Brain and consciousness are comparable to the second car model and not the first. To prove my point, you can look for anatomical dissection models[84] which shows the optic nerve connecting the eye to the brain has a functional afferent fibers, sending signal to the brain, while efferent fibers which sends signal from the brain to the eye,

---

[84] Optic nerve. Author: Lorenzo Crumbie, MBBS, BSc. Reviewer: Elizabeth O. Johnson, PhD. Last reviewed: August 16, 2023

https://www.kenhub.com/en/library/anatomy/the-optic-nerve

SCIENCE AND FAITH

are identified but have no function in vision. Thus, the eye doesn't receive any signal from brain regarding vision, yet all humans feel they are seeing with their eyes.

The relevance of the example in the second model, with the relay of an image on a screen, is how are errors formed. Starting from errors in visual perception, then image relay and image processing, which then gets influenced by the processor (brain) by attempting to fill missing details, and recall past experience and verify the image details with our beliefs, which can be distorted by cognitive bias. This is exactly what happens to our perception and we need to be aware of it, if we are to avoid wrong beliefs. A simple validation of this: is the challenge we experience when correcting spelling errors when reviewing a document, as the brain corrects visible errors in the image based on memory and previous experiences, and then presents it in a correct form to our consciousness, making these errors difficult to spot. The second point of relevance is, it abolishes the first person perspective; thus, anatomically, the first person perspective never exists.[85]

We are born with a clean, empty consciousness, with some inborn traits and characters that influence us; we have some ability to reorient our consciousness to modify these characters, whether they were built-in (i.e., born with us) or even the ones we acquire later

---

[85] The point of view of a story determines who is telling it and the narrator's relationship to the characters in the story. In the first person point of view, the narrator is a character in the story telling it from their perspective. In the third person point of view, the narrator is not part of the story and the characters never acknowledge the narrator's presence. In the second person point of view, the reader is part of the story. The narrator describes the reader's actions, thoughts, and background using "you."

https://www.merriam-webster.com/words-at-play/point-of-view-first-second-third-person-difference in life. We then start carving pathways of beliefs and knowledge in this consciousness; some of these are stored for later utilization, and others are projected to us in our access consciousness. These self-carved pathways could be right or wrong, as we are influenced by many factors which can skew our perception.

An individual who never questions his rectitude is one who tends to fall into false believes or believe first and then get evidence later, or in plain English, is likely to be wrong.

We have limitations and boundaries installed in our consciousness, which saves us from unnecessary realizations, making us freer to concentrate on important matters in our life. These boundaries and limitations probably do not only exist in our consciousness but in our soul and spirit end as well.

The HRV could be direct evidence of the realization of consciousness outside the brain.

There remains another psychological attribute relevant to our Science & Faith topic, which is born with the human being and accompanies him till his death, this is "Fear," and will be discussed in chapter V.

We have come to the end of the design section, where we have gone through some scientific facts and used cross-sectional comparison at different levels of human design. Some of this analogy is a bit philosophical. However, I hope the reader will find it difficult to disagree with most of what is being said.

The reader at this stage should start agreeing in principle with the basic foundation of intelligent design; if not yet, I have one more example to demonstrate, then I suppose if the intelligent design concept is still not settled, there is no reason to proceed any further.

Imagine a Bedouin walking in the middle of a desert and suddenly is faced with a massive castle, which is the same color as the desert sand and is made from the same sand material; the castle is complete with windows, doors and rooms. He would immediately conclude; somebody built this castle from the material available. And he would start appreciating and judging this design as good or bad.

The possibility of this specific design of a castle being created by the wind blowing and natural erosion can be immediately refuted, as a Bedouin would tell you, the latter would carve caverns and caves rather than a specific design with doors and windows. Similarly, in a human design, a process of natural selection or chance would have created a thick cornified material at the end of our fingers instead of a nail bed system on one side and a globally unique identification system in the form of a fingerprint on the other side. A human with vision and, hearing and consciousness is a specific design and not a self-made cavern or cave.

II

Thinking And Reasoning

Topics discussed thus far belong to the science of Physics and Biology, and now other tools need to be employed in the line of reasoning; these are the sciences of Logic, Thinking, and Philosophy.

As we will be attempting to argue for and against many issues, it is befitting we try and use formal scientific tools that are agreed upon. Thus, I thought that brushing on the ways humans think and argue and the systematic way to do so will be appropriate.

Irking the reader's mind with this information might serve a strategic purpose. Not only will it help identify a plausible logic for a correct path of belief, but as important will help him understand how humans have, do and will fall into an opposing or wrong reasoning.

Humans early on, have realized that their perception of events and problems with their consciousness have set them on a conflicting path against each other; and as they developed more social complexities, they needed to create ways to standardize their understanding of social dynamics, ethics, governance, and interpersonal relationships. Ancient civilizations, mainly Greeks (Aristotle), began to systematize ways of thinking to distinguish between sound and unsound arguments. This formal logic became the foundation for rigorous thinking in many disciplines, from mathematics to law.

THINKING VERSUS LOGIC

In 2011, Richard Arum and Josipa Roksa published a book titled: Academically Adrift: [86] The book highlighted an important problem with higher education, where they identified extremely low levels of learning, as measured in terms of critical thinking, complex problem solving, and communication.

According to this study, university students possess weak critical thinking skills. They often respond to test questions by simply regurgitating the information they learned, without demonstrating the ability to analyze or critically understand it. The study confirms that these issues impact the students' ability to graduate from university and succeed in their professional lives in a world that demands critical thinking, problem-solving, and effective communication skills. I don't believe scrutinizing the relationship between science and faith would be any different.

The first time I came across the subject of logic was during completing my surgical training; my main purpose was to see if logic could prove points of argument. I was fascinated by the scientific attempt to classify statements and establish the validity of premises. Later, I faced a slight dilemma: logic can't explain everything in life. An example is consistency, as logic is about consistency. If you're inconsistent, then you're not logical. Let me quote the scenario I first read in a book about elementary logic. When you tell a child don't open the fridge door, then a few minutes later you open it yourself, despite this being the right thing to do, it is inconsistent and thus, not logical. I later learned that logic is concerned with evaluating the structure of arguments and their validity/form. As to the truth of the claims themselves, logic has little to add.

---

[86] https://press.uchicago.edu/ucp/books/book/chicago/A/bo10327226.html

Then I came across the late author Edward De Bono[87] and a few of his published books about thinking. He promoted thinking as a science which ought to be taught comparable to other scientific disciplines. He is the one who coined the term "Lateral Thinking." Indeed, thinking can overcome some rigid laws of logic when discussing a topic so vast and so variable as faith.

Logic is often perceived as cold and calculating, while "Thinking" can be logical but includes other factors such as emotions, imagination, culture, language and social conventions.[88]

In summary, Logic is defined as: *"The philosophical science and art that analyses arguments and inferences to discern valid from invalid forms of reasoning."*[89]

On the other hand, Thinking is: *"The intellectually disciplined process of actively and skillfully conceptualizing, applying, analyzing, synthesizing, or evaluating information gathered from, or generated by, observation, experience, reflection, reasoning, or communication, as a guide to belief and action."* [90]

As much as possible, I will stick to true or false statements. In other areas, if I use subjective or analytical statements, I will declare this so the reader can decide whether to agree or disagree about the statements himself.

---

[87] https://www.debono.com

[88] John Spacey, in an Article titled "Rational vs Logic," What is the difference? October 2015.

https://simplicable.com/new/rational-thought

[89] https://www.ourhappyschool.com/philosophy/logic-critical-thinking-and-philosophy

[90] https://www.criticalthinking.org/pages/defining-critical-thinking/766

THINKING SKILLS AND PHILOSOPHICAL REASONING

How can the human opinion of their Designer be so vastly different, despite having the same data input from evidence of design? Most people would claim it is the result of being born into a faith, which is largely correct, but how did it develop so differently in the first place? Also, with human interaction, there have been many inter-faith discussions resulting in a large amount of migration into and out of faiths. Could other factors influence those who are looking for the truth? Like using different ways of deliberations and interpretation of the data available.

Using science to look inside the human mind to understand how it functions might help us explain the phenomenon of faith variation and identify whether the correct scientific tools were used to find the truth. This will fall under thinking skills and philosophical reasoning.

THINKING SKILLS

Scientists have identified many types of thinking skills; some list four skills, others identify seven types, and some even record up to 19 categories. These skills are thought to be essential for humans to help them understand the world around them, solve problems, make logical choices and develop their values and beliefs. Every individual tends to rely on a type of thinking skill depending on his personality, upbringing, or training. It is believed learning new thinking skills can improve the way humans perform in life. Thus, the type of skill they need to learn depends on what they wish to do.

Types of Basic Thinking Skills

Abstract thinkers: can relate seemingly random things with each other; they make the connections others find difficult to see.

Analytical thinkers: separate a topic into its basic elements and examine them and their relationships. They have a structured way of approaching problems; they use logic rather than emotion.

Critical thinkers: are deeper into analysis and exercise careful evaluation or judgment to determine the authenticity, accuracy, worth, validity, or value of something. Rather than strictly breaking down the information, critical thinking explores other elements that could influence conclusions.

Creative thinkers: will develop ingenious solutions to solve dilemmas and problems in life. They break away from the traditions and norms of society.

Concrete thinkers: like hard facts, figures, and statistics; they edge away from philosophy or abstract thinking.

Convergent thinking: is a process of combining a finite number of perspectives or ideas to find a single solution. They target these possibilities or converge them inwards to come up with a solution. The exercise of this book is probably all about convergent thinking.

Divergent thinking: is a way of exploring multiple solutions to find one that is effective. They go as far and wide as necessary and move outwards in search of solutions.

And more: I would like to add one more skill not usually mentioned in many psychology texts, despite being used since very ancient times by Stoic philosophers like Marcus Aurelius, Seneca, where they regularly conducted an exercise known as a *premeditatio malorum*, which translates as *"premeditation of evils."* [91] This is now considered a crucial skill for product managers and can serve as a very important tool in our topic of discussion here. This is dubbed as an "Inversion thinking skill," and it works by imaging the opposite of the situation you are navigating or the problem you are trying to solve. Apparently, this is how very intelligent and very successful people think. [92]

So, in a project plan, instead of thinking about how to succeed, you would think of how this project would fail, and once you have identified the failure scenario, you work on avoiding it. [93] In simple terms, as an alternative of focusing on what you want to plan, you concentrate on what will sabotage your plan and try and avoid it.

The Thinking Trap

Let us identify the types of thinking skills we will need in our discussion here and which ones we have to restrict so we don't fall into a thinking trap. The idea should be clear. Otherwise, a reader may think I am implying faith is contradictory to thinking.

It is a valuable exercise for an individual to stop and analyze the

---

[91] Inversion: The Crucial Thinking Skill Nobody Ever Taught You. Written by James Clear on his website.

He writes: Inversion can be particularly useful for challenging your own beliefs. It forces you to treat your decisions like a court of law. In court, the jury has to listen to both sides of the argument before making up their mind. Inversion helps you do something similar. What if the evidence disconfirmed what you believe? What if you tried to destroy the views you cherish? Inversion prevents you from making up your mind after your first conclusion. It is a way to counteract the gravitational pull of confirmation bias.

https://jamesclear.com/inversion

[92] Inversion: The Billionaire Thinking Skill You Were Never Taught in School.

https://www.mayooshin.com/inversion-charlie-munger-billionaire-thinking/

[93] https://www.modelthinkers.com/mental-model/inversion

dominant type of thinking skills he uses in everyday life and if he is applying the right skill to the relevant task. An example is an artificial intelligence engineer working on a video game project will need creative thinking. However, a geologist working on fossil evidence for human existence will need concrete, analytical and convergent thinking.

When discussing ideas of Faith and Deism, a human brain should not fall into the trap of using the wrong tools, which will set him on the wrong track; this might lead to more confusion, as we can see happening in the world throughout the last few thousand years.

As a hypothetical exercise, imagine in the first example, the engineer started using concrete thinking to create a video game; he is unlikely to develop an innovative product. Similarly, if the geologist started using creative thinking to try and find facts and evidence of human existence, he could start believing he is the designer of the universe.

Returning to our topic of faith and science, we need to find facts and not shop for an idea of a God or decide on the best offer. Thus, abstract,[94] analytical and convergent thinking will be the best types to employ in this exercise.

It is then possible, a lot of the confusion and differences in human belief are caused by humans applying their inherent or dominant thinking skills but to the wrong task.

PHILOSOPHICAL REASONING

Philosophers have been at the forefront of discussing faith. One arm of philosophy would be very handy, which we will discuss

---

[94] Abstract thinking has many aspects, some of which can be erroneous, as will be explained later.

below. This is logical reasoning and its types. However, according to Leonard Peikoff, philosophy can often be disassociated from real life,[95] as philosophers have an inherent feeling that they need to create new philosophical ideas that don't yet exist.[96] The latter we can label as a philosophical trap.

Logical reasoning, on the other hand, is one of the tools we expect our brain to use; thus, we need to identify and agree about the definition and facts about this tool.

It can be classified into three types:

Deductive reasoning[97]: In logic, this determines whether the truth of a conclusion can be determined for that rule based solely on the truth of the premises. The example mentioned in Wikipedia is, "When it rains, things outside get wet. The grass is outside, therefore: when it rains, the grass gets wet." Mathematical logic and philosophical logic are commonly associated with this type of reasoning. In research, this method will go from the general to the specific, as follows:

Inductive reasoning[98]: This method will go the opposite way from a specific observation to a broader theory. It involves searching for

---

[95] Leonard Peikoff argues these attitudes stem from mistaken views about the basic relation between ideas and reality and that to understand objectivism and philosophy more generally, one needs a proper method of keeping abstract philosophical ideas tied to reality.
https://courses.aynrand.org/campus-courses/understanding-objectivism/philosophy-versus-life-three-arguments-against-philosophy/

[96] Transcription of Leonard Peikoff podcast.
https://curi.us/1807-leonard-peikoff-says-hes-not-a-philosopher

[97] https://conjointly.com/kb/deduction-and-induction/?search=abductive%20reasoning

[98] https://conjointly.com/kb/deduction-and-induction/?search=abductive%20reasoning

patterns from an observation and developing explanations. This type of reasoning is commonly associated with generalization from empirical evidence. Although these arguments are not deductively valid initially, they can be at the end.

Observation ⟹ Pattern ⟹ Tentative Hypothesis ⟹ Theory.

Abductive reasoning[99]: It is called inference to the best explanation, which is the procedure of choosing the hypothesis or theory that best explains the available data. Example: "When it rains, the grass gets wet. The grass is wet. Therefore, it might have rained." This kind of reasoning can be used to develop a hypothesis, which in turn can be tested by additional reasoning or data.

Specific Observation ⟹ Specific Conclusion (May be true).

Although theories and ideas can be a combination of three methods, we should try and adhere to the deductive approach less with the abductive as much as possible, with proper labelling so the reader can follow the logical evidence.

COGNITIVE BIAS VS. LOGICAL FALLACY

Other types of argumentations can be deleterious and potentially corruptive, should be predicted and kept away, lest they get into the thinking process and deviate from the logical trail, these are:

Cognitive Bias: This is an erroneous way of thinking that occurs in people's perception of the data they are processing and interpreting from the world around them. Everybody can fall into this deception;

---

[99] https://en.wikipedia.org/wiki/Abductive_reasoning

one can easily spot it in other people's speech, but it can also creep into one's thinking. It simply means your brain is choosing an easy way out.

The list of cognitive biases is evolving, with almost 200 types identified.[100] Some examples relevant to our topic are:

Confirmation Bias: This is favoring information which conforms to your existing beliefs and discounting evidence which does not conform.

This is sometimes dubbed "my side bias." I have always thought and read comments stating; only intelligent people are capable of changing their minds when presented with facts that are contradictory to their beliefs. However, on searching for evidence in the scientific literature, I was surprised to learn this is just not true; Stanovich et al.,[101] published a study examining the relationship and reviewing previous research done on the topic and concluded *"The magnitude of the myside bias shows very little relation to intelligence. Avoiding myside bias is thus one rational thinking skill that is not assessed by intelligence tests or even indirectly indexed through its correlation with cognitive ability measures."* They discuss "Rational Thought" as a wider concept and skill that can include many other balanced thinking parameters than mere "Intelligence."

The significance of the above-mentioned research is it can explain observable phenomena we see in many communities around the

---

[100] Every Single Cognitive Bias in One Infographic. By Jeff Desjardins. Published by Visual Capitalist. August, 2021

https://www.visualcapitalist.com/every-single-cognitive-bias/

[101] Myside Bias, Rational Thinking, and Intelligence. K. Stanovich, R. West, and M Toplak. Current Directions in Psychological Science. Volume 22, Issue 4.

https://doi.org/10.117 7/0963721413480

world. I once watched a TV program about a professional man who created a God figure for his company and he was recorded praying to the company statue at home. I am quite certain this individual does not lack intelligence.

Causal Uncertainty: When people are trying to find an explanation for why something happened or exists, this activates a cognitive process mainly through abstract thinking. Although some of the top argumentation processes presented in the literature on the topic "science and faith" can be classified as "abstract thinking," which is a higher order of thinking skills, research suggests "abstract thinking" can sometimes promote different types of bias. As people seek to understand events and how everything relates to everything else, it can sometimes cause people to seek out patterns, themes, and relationships that may not exist. J. Namkoong from the University of Nevada, Reno, and M. Henderson, from the University of Texas, Austin, write: *"It is possible the attempt to reduce causal uncertainty through abstraction and simplified causal explanations leads people to hold more extreme views. If this results in ideological, political, or religious extremism, it may affect behavior such as social engagement."* An example in our science and faith topic is when people attempt to explain the observation of similarity in species anatomy and shape; they try to create connections through abstraction, resulting in a hypothesis of a string of evolutionary scales. Or simply that our grandfathers were apes.

Logical fallacy: Is an error in reasoning which makes an argument less effective or convincing. There are many subtypes and classifications of logical fallacies, but the two that are relevant to our debate are the *"Ad Populum Fallacy"* (The Appeal to Popular Opinion Fallacy); this is when one argues something is true, just because a large number of people or influential persons are doing it or believe in it. And "The fallacy ad ignorantiam"[102] (Appeal to Ignorance Fallacy) is when one asserts that something must be either true or false because it hasn't been proven to be one way or the other. In other words, a particular belief is said to be true because you do not know it is not true. We will refer to this example later.

Now, you should fasten your seat belt as things are about to get more exciting when we start analyzing the topics mentioned above and see what we can learn about our "Design" and our "Designer" using logical reasoning.

---

[102] https://fallacyinlogic.com/appeal-to-ignorance/

III

On The Origin Of
a human...

By trying to apply all the above scientific data mentioned so far and inferring this to some observations of human life, we can make some analytical theories of our existence and question some of the existing ones.

HUMAN INCARCERATION THEORY

One of the first theorems I would like to propose is the "Human Incarceration Theory." Which demonstrates how much limitation is applied to the human race.

The universe we live in is probably made out of trillions upon trillions of planets, if not more, while the "Human Race" seems to be stuck in only one.[103]

This is like a group of people living in a massive castle, where they are born, live and die in one room; they have managed to take some pictures of other rooms of the castle from the window and visit one or two of the closest rooms. Nevertheless, they are still hoping to meet others on the higher floors who might or might not be there.

The analogy is meant to prove my proposal we humans are

---

[103] Cosmic Eye — Universe Size Comparison Is a video published on YouTube in May 2018. https://youtu.be/8Are9dDbW24?si=T-qIGwSjTxY3eXho

This is the original landscape-format version of the short movie Cosmic Eye, designed by astrophysicist Danail Obreschkow. The movie zooms through all well-known scales of the universe, from minuscule elementary particles out to the gigantic cosmic web.

incarcerated and locked up in a very limited space compared to what is available.

This brings to mind the famous Pale Blue Dot speech by Carl Sagan, who, on February 14, 1990, requested NASA to re-direct "Voyager I" around to take a photo of Planet Earth.[104] It took a picture of the Earth from 6 billion kilometers (3.7 billion miles) away. Carl Sagan then called it: *"that tiny pale blue dot"* or *"a mole of dust suspended in a sunbeam."* Scientists believe it to be one of the most powerful photos ever taken. Excerpts from his famous speech given at Cornell University in 1994 read: *"That's us. On it, everyone you love, everyone you know, everyone you ever heard of, every human being who ever was, lived out their lives." "The Earth is a very small stage in a vast cosmic arena. Think of the rivers of blood spilled by all those Generals and Emperors so in glory and triumph, they could become the momentary masters of a fraction of a dot."*

On a similar analogy, try and look at planet Earth pictures in cosmic images and imagine what Carl Sagan said. You can soon logically deduct that the entire human race from the beginning of its inception to date, whether living or dead, is on this small blue planet.

And if that is not enough, you can look at our reach capabilities on our planet. We can fly that far high in the sky with our greatest and most powerful inventions and so little down into our Earth's crust.

The limitation doesn't stop here; our visual ability is inferior to most mammals, our hearing powers are very limited, and our strength to run is mediocre.

---

[104] https://www.planetary.org/worlds/pale-blue-dot

Perhaps the most compelling boundary imposed on the human race is the "Linearity of Time." Humans define time as a series of now moments; we consciously call some of these now moments "Past" and the ones we are waiting for "Future." All these now moments keep moving on. We don't seem to be able to get past its boundaries or stop its progress as it appears to be taking us to a pre-set destiny. Time travel is an attempt by humans to imagine how life will be like if we are freed from its constraints. So far, it seems like wishful thinking. To demonstrate this point, it would be appropriate to mention the experiment Professor Steve Hawking did on people who believe and claim they can time travel. When he set up a party but sent them the invitation after the party was over, he then decided time travel is not possible when no one could travel to the past and attend.[105]

Why then, was it decided that we are locked up with such limited reach? The answer is I don't know, but it would be foolish to presume this was another simple chance and not part of the integrated intelligent design we are trying to explore.

COLLECTIVE SCOPAESTHESIA

Scopaesthesia is defined as a psychic staring effect,[106] a phenomenon in which humans detect being stared at by extrasensory means. Many humans reported this feeling, which has attracted little research since

---

[105] In a Discovery Channel series, "Into the Universe," Professor Hawking was a theoretical physicist, cosmologist, and author who was director of research at the Centre for Theoretical Cosmology at the University of Cambridge. Between 1979 and 2009, he was the Lucasian Professor of Mathematics at the University of Cambridge, widely viewed as one of the most prestigious academic posts in the world.

[106] https://www.skepdic.com/staringeffect.html

1898 by E Titchener. Some studies explained this by a false feeling of being watched; others utilized child psychology to rationalize the intuition, where we grow up feeling a watchful eye on us when we make every move. Naturally, this is by our caregivers. And we keep Scopaesthesia all over our life, some more than others.

Collective Scopaesthesia is a theorem I am proposing that poses a question: What if we, "the human race," are being watched? Are there any inferences we can use for this abductive hypothesis?

A. The feeling of being watched is a common phenomenon for individual humans, with some reporting it more than others, but it is often conveyed without explanation.

B. Randomised trials have failed to explain its occurance by looking at other fellow humans watching test subjects, but what if a much higher power is collectively watching us?

C. It is logical to presume that a very complex, dynamic, changing universe, has a designer who would be interested in watching over his design. Once we accept the existence of a universe that can't make itself but was made by an intelligent design power, we can then conclude this power would be interested in watching over their design.

D. Chaos Theory and the Butterfly Effect: Looking at these two phenomena and making connections between them, first is the "Chaos Theory," which was first discussed in 1880 by French mathematician Henri Poincaré.[107] To explain the point I want to make, let us look at the simple human mind

---

[107] https://thepangean.com/The-Order-in-Chaos

and how much it likes organized behaviour in objects, as this helps the simple computational brain to deduct reason; once this starts to become complex, this behaviour appears unorganized and the human brain gets confused and labels the behaviour as chaotic (unorganized), or is it? Mathematicians realized things that look chaotic could and do have some kind of repetitiveness, but due to their complexity, we cannot perceive their underlying patterns.

A simple example is weather systems, which look chaotic to the observer, but the main computer frames can make computational sense out of them, which can lead to accurate future predictions. The same can be applied to human behaviour, traffic control and stock market predictions.

Second is the *Butterfly Effect:* Starting in 1800, Johann Gottlieb Fichte says[108] *"You could not remove a single grain of sand from its place without thereby changing something throughout all parts of the immeasurable whole."* Later, this was put as a theory in 1963 by Lorenz, who published a theoretical study of this effect; he documented running a numerical computer model to redo a weather prediction. He entered the initial condition 0.506 from the printout instead of entering the full precision 0.506127 value. The result was a completely different weather scenario.

By combining the two theories together, one can claim the universe is running with an organized chaotic pattern

---

[108] https://www.grisda.org/cambrian-explosion-and-darwins-doubt-1#:~:text= The% 20pattern %20of%20abrupt%20appearances%20in%20the%20Cambrian,that%20 new%20types%20of%20organisms%20could%20be%20recognized.

(sometimes dubbed *Quantum Chaos*)[109] with the possibility of a butterfly effect which could lead to things getting out of proportion. The human race has probably inflicted a major butterfly effect on their surroundings and close environment. This leads to the question: why humans have not annihilated the universe with their destructive methods? And the most logical answer would be, something or somebody is watching them. After all, the forces humans are tapping into are formidable.

E. Einstein was a proponent of a *"deterministic nature of the universe,"*[110] and objected to Max Born's publication in 1926 of the *"probabilistic chances"* of the quantum universe. His view was the universe is like a great and complex clock, simply ticking forward in a complicated but entirely predictable manner. Einstein subscribed to this vision and used his most famous quote, *"God does not play dice with the universe."* This is synonymous to watching over, and was not any religious belief, but a scientific abduction from a great scientist.

F. In an open-ended question, *"The Universe Should Not Actually Exist, CERN Scientists Discover."*[111] This headline was reported in all news media in October 2017. Scientists working at CERN, the European Organization for Nuclear Research, have said that the universe as we know it should not exist. After performing the most precise experiments on

---

[109] Quantum chaos is a branch of physics which studies how chaotic classical dynamical systems can be described in terms of quantum theory. The primary question that quantum chaos seeks to answer is: What is the relationship between quantum mechanics and classical chaos? https://en.wikipedia.org/wiki/Quantum_chaos

[110] https://www.wondriumdaily.com/quantum-universe-fundamentally-probabilistic-not-deterministic/

[111] https://www.newsweek.com/universe-should-not-exist-cern-scientists-discover-692500

antiprotons that have ever been carried out, researchers have discovered a symmetry in nature they say just shouldn't be possible. The findings from the BASE (Baryon Antibaryon Symmetry Experiment) are published in the journal "Nature".[112]

First author Christian Smorra, from Japan's RIKEN institute, said in a statement. *"All of our observations find a complete symmetry between matter and antimatter, which is why the universe should not actually exist."*

This scientific publication is saying that according to science, this universe should collapse from the scientific point of view and something is holding it together. We can pose this question to many people: Who or what is holding the universe from collapse? Science can't explain this so far.

I am quite sure if you ask followers of the 50 thousand religious sects from around the world, their answer would be, our God and the High priest are doing it together.

My comment to the religious sects would be, "It's too late." If you could have declared this information before scientists could find it, maybe we could have believed you. Maybe we should meet in the next book if you have already done so.

My Statement on whether we are being watched: Here, we are reaching the edge of science; maybe one day, scientists can find a scientific explanation about the feeling of being watched. Or perhaps we are at the end of our universe's physical forces and starting into metaphysics. Or maybe somebody is watching us and holding the universe from collapsing.

---

[112] Q Smorra, C., Sellner, S., Borchert, M. et al. A parts-per-billion measurement of the antiproton magnetic moment. Nature 550, 371–374. 2017.

ATRII SCRUTATIONIS (EXAMINATION HALL)

I have spent most of my life studying and sitting exams, and now I still go to the examination hall again as an examiner. And although the last exam I had to take was a long time ago, I still have a constant recurring dream that I have a physics exam the next day, for which I am not adequately prepared. Is this why I think we are in an examination hall? Or my brain can identify the similarities and the premises.

The first resemblance between this universe and the examination hall is in both scenarios, once you leave, you don't get to come back, and the logic usually is, to protect the exam's integrity.

The second is the absence of a clear signature on the universe's design. It is evident human beings like us, as soon as they start their journey in this life, will be asking questions: first of all, who made this? The absence of an explicit declaration indicates we are meant to look for it.

"An exercise in which an answer is missing" is either a challenge or an exam. As we are obviously no match for the Intelligent force behind this design, it's not the first; thus, it must be the latter.

The third similitude is the "Human Identification System." We have always received unique exam identification numbers, badges and recently, barcodes. The idea is to ensure the correct person is sitting the precise exam. The design process comparably, has given all humans alive or dead, from the beginning of humanity till its end unique identifiers, like DNA sequence, iris image of the eye, and fingerprints. Obviously, we humans are being labelled, marked and tracked very precisely.

Taking the fingerprint as an example: In 1788, a scientist J.C.A. Mayers declared "patterns and arrangement of skin ridges in humans are never duplicated in two persons." Later, in 1880, a Physician, Henry Faulds, noticed the same phenomenon and later examined children's development over two years; Faulds found their fingerprints stayed the same. He wrote a letter to the journal "Nature" arguing that fingerprints could be a way for police to deduce identity.

If we can imagine an unintentional, undirected process of creation, there would be no logic to come up with such a sophisticated and reliable identification, as this is not linked to survival as evolutionists might argue.

The question is, how many other identification systems do we have within our current design, and whether our souls have any comparable identifiers?

The fourth and most profound premise in the examination hall hypothesis lies in the issue we discussed earlier, which is "Fairness." The Designer has shown the ability to create closed, faultless systems, demonstrating different types of fairness. These are absolute in precision yet variable and adapting. I urge the reader to go back to the section on fairness among atoms, human cells, and human organs. If one scrutinizes the universe and the physical laws we discussed earlier, one can conclude promptly, the designer of all this loves fairness.

Why, then, would the designer decide to leave fairness at the level of the individual human or the human race unattended? A common layman's phrase "life is not fair" could be easily described as a logically true statement. But a better accurate question to pose is:

Are humans and the human race fair? After all, we agreed they are the only section of the universe design that was given the power to act with a will (remember this is not the same as free will, so we don't lose the readers who don't believe in free will).

The answer to the question, using our previous argument, is "Humans exercise fairness, some of them, some of the time."

Fairness Dichotomy

It seems the design process of this universe has decided on absolute fairness and rigid non-choice options at the levels of atoms, biological cells and human organs and then moved to the "Act with a Will" option for the Human and Human race levels.

We can then deduct, it's the freedom to act with a will that was conferred upon the individual human and the human race as a whole, which is responsible for the unfairness we see in our universe."

The dilemma is: What are the consequences of this unfairness, or will there ever be one?

Alternatives to Human Unfairness

Why, then, is "Unfairness" left unattended and is this a design fault? To put this question into a conditional argument, utilizing the "Inversion Thinking" tool we discussed earlier, by looking at the "What-if Scenario," What if there was absolute fairness between humans and the human race, and everything was good? And here, there would be two worlds we could envisage.

Universe A: What if the design process involved our fellow humans and us being as fair as a clock or a proton in an iron atom? Without any kind of "Will" on our part.

Consequences:

    a. Life would be extremely boring, as we would lose our Free Will, something which humans are very proud of.

    b. We would not need much computational brain power.

    c. We would probably not be interested in the design process.

    d. We would not be having this conversation.

    e. The reader can then ask himself, would he have wanted to live in a world like this?

Universe B: What if we still kept our freedom to act with a will, yet consistently "Good" wins over "Evil" and justice is achieved in front of our own eyes? It is necessary here to mention, I have failed to find a logical or consistent definition of good or evil. It is fair to say this is one of the most controversial issues in human history, probably more so than religion, as even among the same religion and even within the same sect, people will differ about what is good and what is evil. Moreover, the perception of good and evil has changed over human history many times and what was acceptable in the Roman Empire would probably send you to prison today. We can abduct our understanding of the topic will continue evolving in the future as well.

Consequences:

    a. As there is no method to define good and evil, this is impossible to achieve unless all humans agree on one set of laws, religion, or code, which has never and will never happen in this world. Otherwise, they will perceive unfairness from the retribution falling upon them, which will lead to more actions of wrongdoing on their part. We could probably lose our logic tool, and life will not make sense anymore.

b. "Direct justice" and "Freedom to Act with a Will" are both incompatible to exist at the same time.

c. If evil then loses constantly in front of our eyes, the percussion of constant punishment and reckoning against evil-doers would make life unbearable. And probably would self-destruct.

d. If we constantly view direct retribution, we would clearly avoid evil, and the result would be universe B changes to universe A, and the preceding consequences prevail.

Further premises that will enforce the examination hall hypotheses will be presented later during the discussion of "Human Fear," "Human Disease," and "Human Death."

The Examination Process

Any exam scenario needs an Examiner, Candidates, i.e., examinees, as well as the Examination process. What information can we deduct about this examination process if it exists? The only logical deduction is, it must be proportional to the amount of relative freedom we have been given in our lives. I.e., if you are chief of a tribe and have all the resources available to you, you cannot have the same liability as a poor old woman living in a shabby hut looking after an ill husband. Another obvious factor would be the amount of damage you managed to inflict during your lifetime and an example is if you were a horrible person to your colleagues at work, it might not be very pleasant for you, but if you were atrocious to an entire nation or race you probably will be assessed on a different level.

Good vs Bad Exams: When we trained in the examination process, we learned questions which are answered correctly by everybody or those that not answered by anybody are both bad and ought to be removed from the question bank.

The relevance of this to our discussion is a well-thought-of design will probably have a well-orchestrated exam. Thus, we should not expect the questions we face in life or in any coming meta-universe to be so easy that we all can comprehend and answer them well or they will be so incomprehensible nobody can understand. Rather, there will be constant discussions and deliberation within our own brains and among ourselves.

Thus, the presence of constant debate, as well as the time and space available for us to comprehend, analyze, make up our minds and lay down our believes eloquently, proves this is a well-planned exam process.

We can then subjectively conclude the section of the examination hall scenario by claiming the room we are living in (Planet Earth) within this massive castle (Universe) is literally an "Examination Hall." We are writing our believes and answers to the problems we face daily on this earth; this hasn't been marked yet. So, we are not exactly sure what is right and what is wrong. By allowing humans to act with a will and give them a chance where they would struggle to support good and fight evil, and when they depart from this universe, they will be transported into a "Meta-Universe," where there is a measuring stick, or scale which they would agree upon.

This will then give them the exam results of their achievements. Then and only then this ultimate design will be complete.

Some of us are trying hard, and others don't care (they are quite happy to leave the examination answer book blank). But the marking of this exam is coming, and as we haven't seen it yet, it must be in a meta-universe or an alternate universe.

To me, "planet Earth" looks like an examination hall and certainly feels like one. Remember the famous English phrase, "If it looks like a duck and walks like a duck, it must be a duck."

Boundaries of Science

Does everything in this universe follow the rules of science, i.e., the four physical forces? Depending on whom you ask, the reply will differ. For example, naturalists will tell you yes, while theologians will tell you no. Our discussion linking science and faith will not be complete without putting this debate to a logical argument.

I would like to refer here to the story of Sir Isaac Newton and the apple. I remember hearing the story of "Newton and the Apple" in fourth grade, and I have been captivated ever since. I recall returning home that day, thinking if Newton became great by noticing such a very trivial observation, then there must be many others I could come up with myself. The funny thing is my child-like mind said what about an orange, then I quickly realized it was the principle of gravity we were talking about here and not the object of the apple. For many years, I struggled to find another comparable "obvious thing which could lead to something unobvious," but I failed.

Then, in my early twenties, it hit me. Why would Newton be surprised by the apple falling down and not up? Yet he was not surprised by the existence of the tree itself. After all, the tree is a live being, grows from a seedling, can repair itself if injured, is constantly growing, and produces fruits, i.e., functional. How could a tree then be called "just a tree," and all it does is perfectly logical? So, Newton recognized the apple but missed the tree; by noticing the apple falling, he discovered "Gravity," however, if you appreciate the tree, you will ascertain the "Designer of Gravity."

The brain that can look through the obvious and not only at it, subtract the usual, and deduct underlying facts is a brilliant mind. When I came across some of Edward de Bono's writings about thinking skills, he explained the mechanism of the trap all humans fall into and how to free your mind and bring out the Newton within you.

Edward de Bono writes, *"Because of the way the nerves interact with each other in the brain, there is a dominant track that will suppress any alternative tracks. Access to the side track is impossible. If, somehow, we eventually manage to get across to the side track, then in hindsight, the connection back to the starting point is easy and obvious. This side-track thinking is what gives rise to creativity."* He then adds to this by questioning: *"How do we make the lateral move? That is where the deliberate techniques of lateral thinking come into use."* [113]

I think this perfectly explains Newton's ingenuity by inhibiting the common track of thinking and going laterally.

---

[113] https://www.debono.com/teaching-creative-thinking-3

Trying to link this to our discussion of the boundries of science, we need to inhibit the obvious neuronal path of our brain and think laterally. First, let us repose the question again: Which element or structure in this universe does not follow the laws of physics? If such an element or structure exists, it might serve a grand purpose, mainly giving us further evidence a meta-universe exists or is coming. This takes us to the discussion of the "Human Soul."

HUMAN SOUL

Any deliberation on this topic will be out of the scope of elementary logic or physics. Even looking up religious references failed to bring up any exact definition of the human soul. I found some discussions about the difference between the Soul and Spirit, which I initially dismissed as irrelevant, but on a later note, found a scientific logic for its truth; however, we will leave this for a later discussion.

The only knowledge we can congregate here is that the Soul is a specific character instilled in each human, while the spirit is probably a universal core engine driving the soul. Some religious sources claim the soul is eternal and is what carries on living after death while the spirit disappears.

The part of the spirit disappearing could not be further away from the truth if we consider what we have learned from the current design and about the Designer. Precisely nothing in this universe disappears.

Let me give an example from the universe design, which is water. Professor Ken Carslaw, Atmospheric Science, University of Leeds, says in a podcast about water:[114] *"Water is recycled through the water cycle. It evaporates from the oceans, forms clouds, rains (or snow), and the rivers return the water to the ocean. However, in photosynthesis, water is slowly destroyed chemically (plants converting carbon dioxide and water to sugars and oxygen) and recovered again in respiration (basically the reverse of photosynthesis to make energy and CO2). The Earth's plants take up about 12,000 billion kg of water annually. The total water on Earth is about 1400 billion kg. Within about 100 million years, most of the water will have been chemically destroyed. Dinosaurs lived 65 million years ago. You can calculate how much the same water remains from the dinosaur age from the total amount of water on the planet and the amount of water taken up in photosynthesis per year. Thus, some of the water we drink is the same water that was used by the dinosaurs, but more than half is different water."*

In conclusion, things transform from one form to another, and even water never disappears; it is just a different form of water. How could we infer then that the spirit disappears?

In our narration here, we will consider the soul and spirit as one unit and refer to them as the "Human Soul". It was just important to correct an error of logic based on current design principles (that the spirit disappears).

Evidence the Human Soul Exists

The irony is, this is something which is present in every human being. Yet, we can't have any scientific method to prove it exists, and this is precisely what puts it in the category of "Out of Scientific

---

[114] https://www.thenakedscientists.com/articles/questions/drinking-water-dinosaurs-drank

Boundary."[115] If it was bound by any of the four physics forces discussed earlier, it would have been at least detectable, calculable and measurable.

Similar to consciousness, in order to get a deeper understanding about the "Soul," we can look at our status when we lose it. One can see a strange phenomenon when caring for patients in the intensive care unit: a human body functioning with life support, and suddenly, everything stops. Machines and medications can sustain a lifeless body for some time by stimulating the heart to beat and inducing electrical activity, which can provide sustenance to some organs and cells, but there is no life. While these machines and medications can keep a lifeless body functioning for some time, they cannot bring the person back to life. Sometimes, the difference between a body with or without life is just a few seconds apart, but the transformation is staggering. "Something has just left."

This is an experience many humans have gone through with relatives or friends, a Rubicon, which once crossed, all scientific rules and laws don't apply anymore.

Coming back to the topic of whether soul and spirit exist separately or are the same thing. We can hypothetically infer what has left a dyeing human is his soul and spirit. But let us examine another scientific phenomenon, where scientists can grow human cells in a lab. Of course, they are initially removed from an existing live

---

[115] Immanuel Kant said there is a noumenal world that is not subject to scientific laws whereby God, the soul, and morality exist. However, they are postulated with no further proof since we cannot know the world as it is but only as it appears to us (through cognition/physical boundaries). Kant's entire project was to limit science so he could make room/sense of 'freedom'.

https://speedypaper.com/essays/kants-distinction-of-the-noumenal-world-and-the-phenomenal-world person, then immediately inserted in a tissue culture material and supplied with the necessary nutrients. Normal human cells can live and even grow and multiply, then age and die. Other cell lines from tumor cells can be made to live for an extended time.

The above scenario demonstrates these cultured cells are alive and, by default, have a drive to continue their life span, yet they have no individuality of "Will" or "Consciousness." If we stop supplying nutrients to these cells, they will die, and if they just cross this line and we resume nutrition supply, will they come back to life? The answer is no; once a cell has died, it will never come back, even if the elements of life are restored. Thus, it's not a mechanical or physical process. This can make us hypothesize the human cell lines have the universal life core engine (spirit), while only a complete human has the soul.

This probably is good scientific evidence that there is a difference between soul and spirit outside the religious narration.

Another state of humans can be examined when they are in deep sleep or have been anesthetized for operations; we can observe here that they lose consciousness and freedom of choice; even their dreams can be scary and unwanted, yet they can't shut it down, and in this state, we can hypothesis that their spirit is functioning as their body and organs are in full function, but their consciousness and soul are suspended.

Why Can't We Feel Our Soul?

In trying to explain why we cannot feel or realize, our human soul, I have been using the analogy of the "Wired Brain," or lately, "How is it like to feel like a bat?" concept.

By the wired brain, I mean how we are pre-wired to sense different types of fear, empathy, feelings and characters, and in a contradictory way, we are not wired to feel other things.[116] Like how our kidney, liver and other organs function, as we are not wired to do so, and this is a good explanation of the concept.

Throughout the history of humanity, any attempt to feel, capture, or even define our soul has failed. Some thinkers have claimed the soul to be part of the physical process, or a brain function, or it is just energy. Despite that we know these parameters listed above are controllable, either chemically or surgically, more so energy in any form or shape is calculable, cannot disappear, or even feel or have consciousness.

This led me to be persuaded that the human soul is just one more boundary by design which is built in our consciousness.

Then again, if we are sure every one of us has a soul, which leaves our body at some stage in our lives, and inferring to the fact we know about the Designer, nothing of what he designs gets to disappear. As we cannot detect or see any evidence of a soul after its departure in our universe, we can logically conclude: a meta-universe exists and the only thing from this universe which will be passed on to the meta-universe is the human soul.

Meta-Universe: (Does Evil Get to Win?)

Despite the numerous religious rhetoric that good will always prevail, we hardly see this in our lifetime. Rather, a truer statement most rational humans agree about is: There is a constant struggle

---

[116] Refer to the section of consciousness as part of the psychological design in Chapter I.

between good and evil.

However, can we close the above statement at this level? If we do, then we admit evil can, and does, sometimes win.

We can then pose the most crucial, serious, and life-changing question for all humanity: Will the Designer leave his perfect, well thought and articulated design without achieving fairness?

If the answer is "YES he would," then one would fall into the same dilemma as atheists fall into, why would I then be a law-abiding citizen if I am born poor and can't find treatment for my kids while my neighbors are rich and well off? And we have to come up with an explanation: Why would a powerful and intelligent Designer allow evil to win over good as we see in reality, despite having obvious examples of the ability to create fairness?

And if the answer is "NO, he will not leave unfairness unattended," then I think we are heading for a "Major Event."

I just need to clarify, I do not infer, in any way, atheists break the law. Certainly, any sane human being can see the value of being governed by law and can envisage the deleterious effect of breaking the law. However, we do see people breaking the law all the time from all sects of society. Designing a universe where a large number of its inhabitants can break the law, live normally afterwards and eventually get away with it till the end of their lives is not an intelligent parameter. This will be like creating an unspoken law which says: "Do good, and if you don't, just get away with it." Thus, what I am inferring is; an Intelligent Design cannot have an unintelligent parameter, and if you see one, don't rush to a conclusion; wait, you will eventually see balance achieved, as you have already seen in so many design attributes around you.

What Are the Logical Premises That A "Meta-Universe" Exists?

Denying the existence of a Meta-Universe life brings to mind the hypothetical fictional dialogue that was published in an eBook: "Does Mother Exist? Is There Life After Birth?" by Pablo J.Luis Molinero, where two embryos argue during their presence in the womb, and the conversation goes as follows:[117]

One embryo asked the other: "Do you believe in life after delivery?" The other replied: "Why, of course. There has to be something after delivery. Maybe we are here to prepare ourselves for what we will be later."

"Nonsense," said the first. "There is no life after delivery. What kind of life would that be?" The second said, "I don't know, but there will be more light than here. Maybe we will walk with our legs and eat from our mouths. Maybe we will have other senses we can't understand now."

The first replied, "That is absurd. Walking is impossible. And eating with our mouths? Ridiculous! The umbilical cord supplies nutrition and everything we need. Further the umbilical cord is so short. Life after delivery is to be logically excluded."

The second insisted, "Well, I think there is something, and maybe it's different than it is here. Maybe we won't need this physical cord anymore."

The first replied, "Nonsense. And moreover, if there is life, then why has no one ever come back from there? Delivery is the end of life, and in the after-delivery, there is nothing but darkness and silence and oblivion. It takes us nowhere."

---

[117] This fictional story was widely distributed all over the internet,it was published in a book by Dr Wayne W Dyer, January 2008. "Your Sacred Self," however, the detailed search led to a short eBook published by: Pablo J. Luis Molinero. A full document could be found here: https://pablomolinerodotcom.files.wordpress.com/2014/08/boy-and-girl2.pdf.

"Well, I don't know," said the second, "but certainly we will meet Mother, and she will take care of us."

The first replied, "Mother? Do you actually believe in Mother? That's laughable. If Mother exists, then where is she now?"

The second said, "She is all around us. We are surrounded by her. We are of Her. It is in Her that we live. Without Her, this world would not and could not exist."

The first decalares: "Well, I don't see Her, so it is only logical She doesn't exist."

Claiming there is no life after we move along from this universe, would be like using the "Embryo's Logic."

Now to the logical premises a Meta-Universe exists.

Hypothetical Capability: Can you put your hand over your heart and swear, this is impossible, as the designer of this universe cannot do that? If you cannot eliminate the possibility, you must presume it is at least possible. This admission is a great success and the beginning of setting up your examination answer book in the right order.

Note: This does not fall into the logical fallacy type of "appeal to ignorance" we have discussed earlier. As we are inferring to the capabilities we have already seen in the design of humans and the current universe.

Intelligent vs. Complete: In many media coverage of the concept of Intelligent design, one can read the comments on how chaotic the design is, despite being intelligent. Inferring to the unfairness, disease and suffering, this is commonly discussed by some of the great minds and academicians on the topic.

The missing point here is, describing this design in one word as intelligent is not comprehensive, and we are missing another description, which is entirety, totality or completeness. For those who think this design is over when we die. We can ask them on what premises you have based your conclusion on, as we haven't seen Part II yet.

Achievement of Fairness Theory: The most compelling evidence is that an able Designer who have demonstrated extreme ability at fairness and harmony at all levels of the universe design (in some instances, scientist have described it as knife edge precision) except at two levels (the human and human race), will not leave his design incomplete. This means it is not lack of fairness rather a delayed or postponement of fairness. And it would be foolish for a human brain to presume such a powerful force and intellectual designer cannot achieve fairness or will concede to perpetuated unfairness. As inferences from existing designs have shown us, it can be achieved fairly and precisely.

Meta-Universe Logic

Trying to follow up on the "Embryo's Logic" to get some deductive reasoning about what the meta-universe life would be like; which is not religious text, is very hard, but we can try.

Lack of body mass: Life in the meta-universe will have no body mass as we know it now; this is quite obvious as we have proof the current bodies we are wearing gets to disintegrate in infront of our eyes. Whether this will be replaced by another or not is beyond our comprehension.

Soul & Spirit: We have already discussed our human soul; we can presume of its existence, and we can infer it has no mass. Thus, physical restraints are eliminated, like, for example, travel speed; maybe we will get to see the rest of the universe and beyond.

Eternal Life: Within the boundaries of this universe we know now existed for 3.5 billion years, everything is made out of matter and energy and is bound by a linear time frame, i.e., everything created in this universe has a beginning and an end. Outside the dimensions of this universe, this concept does not exist. Thus, our souls will get to live forever in the meta-universe.

Lack of need for sustenance: Take a few moments and contemplate the following scenario: you have departed from this universe and you are now in the meta-universe, you have given up your current physical form and live in Soul and Spirit form, and there is no need for sustenance anymore.

Mentally subtract the effort and work you do for your daily sustenance for yourself and your family, and consider this is self-provided. How will you think then? And how will you re-formulate your opinion about every other human you have known? What would you wish that you had done in this current life?

I will give you some suggestions like: Would you have spent more time with your family? Would you have smiled at your boss in the morning? Or would you have ignored the poor woman standing on the corner of your street asking for help?

The notion is our priorities will change, and the relevance of this point is if we can envisage what our main concerns will be, perhaps we can be more prepared and comfortable with the concept during our life in this universe.

DARWIN'S THEORY OF EVOLUTION

The first time I recall going to the zoo was with my parents when I was six years old. I remember my sudden amazement at the appearance of the chimpanzee peeling a banana, something most probably all of us have experienced. The first idea that struck me was the similarity between us humans and this particular animal. You can call this "The Obvious."

We cannot thoroughly discuss human design without including and exploring the most protected scientific theory in the history of humanity that has attempted to explain our existence. This is likely the only politicized, theologized, and demonized scientific theory to date. To the extent mainstream religion considers it blasphemy and mainstream science does not permit or allow much criticism in this context. Charles Darwin expressed the theory of evolution in his book "On the Origin of Species," published in 1859.

Despite being hypothesized before the discovery of DNA, with the presence of many loopholes admitted by its principal author. However, in the scientific community, it is subtly admitted there are missing links.[118] Yet, the theory is taught and considered mainstream science pending the discovery of the missing links. This has not happened to any scientific theory to date.

The biggest fight is usually between Theologians and Naturalists (as they are called in the scientific communities).

First, I would like to clarify the term, contrary to the common perception, naturalists[119] is not a bad word or meaning; instead, it is

---

[118] The Encyclopaedia Britannica states: "These ancestors have yet to be identified, but ape-hominid divergence may have occurred 6 to 10 million years ago."
https://www.britannica.com/science/missing-link

[119] https://plato.stanford.edu/entries/naturalism/

a philosophical movement which combines philosophy with science and claims all processes in nature must happen within the laws of nature and physics.

Consequently, many religious and common people align themselves with one side, and many scientists who don't necessarily believe in naturalism align themselves with the other group. The irony is the secondary supporters became avid proponents and opponents despite having never delved into the details of the theory or studied the consequences from which it arose.

The theory serves a very imperative purpose: it closes a door by answering an arduous question. "Where did humanity come from?" Which, according to the answer leads to a more serious second question. "Where is it heading?" The elucidation of which could have fearful consequences. If then we postulate, they were created by chance, then there will be no need to investigate the second uncertainty.

Having concluded the same idea as Darwin upon watching the chimpanzee as a child, why do I have some reservations with the Theory of Evolution?

Before commencing our logical argument, I would like to claim Darwin never stated this is antithetical to God's existence; he just attempted to explain how we came to be the way we are. So, hypothetically, God could have chosen to create humans in stages commencing from a Chimpanzee, or he could have chosen to do it the other way, i.e., create each species separately. It's the secondary supporters or the Darwinians, rather than Darwin himself, who have become the main drivers in this movement, as they want to use this concept to prove their agendas. Thus, the theory of evolution does not exclude divine intervention per se.

Humans have always strived to understand how things around them are made. Let us imagine we found a strange-looking machine in the middle of the desert; the first question we will ask is, what is this? And the answer would be a machine. It would be foolish to ask immediately how was it made. The second question would be, what does it do? In the case of a human body analogy, we already know what it can do; the third question that follows is: how does it function? i.e., how does it do the things we see it can do? Once we grab the idea of how it functions, we can contemplate how it was made.

Let us presume we keep finding more and more types of activity and performance in this machine. It would then be premature to ask how it was made, and if the question is posed, someone would definitely say, let us understand its full function; maybe then we can tell how it was made.

Taking the analogy to the Darwin's theory of evolution, how can one man, who had no knowledge of the DNA or microscopic anatomy and physiology of the human body, be able to explain how humans were made? To this day, scientists keep finding multiple layers of functionalities in the human DNA, and sections which were once described as useless are now known to have a function. It is such a long shot for Darwin to achieve that.

But one might ask: What if he was lucky? This is probably the only explanation; he was fortunate as he gave scientists who came after him the excuse and the only way to crack and escape theologians strong grip and influence over humanity for a long time. The aim changed from "Finding the Truth" to "Escaping Theology".

The Observational Argument

I claim Darwin's theory of evolution is an observational theory, as there was no investigation into essential elements like DNA. Thus, an observational argument against an observational theory would be the logical way to start. Here I have a few points scratching my brain:

A. Suppose Species A (Chimpanzee) started, followed by B, C, D, E, and F.....till Species X(Human). How can species A still exist together with species X, without any, even mild existence of B, C, D, E, F...... If the theory was true, we would have seen all those variations living along with us, and their DNA would have allowed them to survive. As it did for A and X, they would have been an improved higher version of A. Most scientists claim only fossil evidence is enough; however, even this is lacking with their own admission.

B. Trying to argue the scenario was not a linear succession, one from the other, rather us and chimpanzee having a common ancestor, does not make much difference, as why did the monkey persist for 15 million years without change while we continued to do so? Even in a tree of evolutionary change, you can have multiple linear pathways; the problem here is the same: we don't see any ancestors of the kind mentioned alive now or the intermediary species that are supposed to be in-between, alive today, as they are supposed to be a better version than A or D.

C. The similarity between humans and other species, whether in shape, behaviour, function or even in our building blocks (DNA), instead of being used to construct a theory of

"Common Origin," could more plausibly construct a theory of a "Common Technique" or a "Common Designer."

D. The Cambrian explosion, Cambrian radiation or early Cambrian diversification: This refers to an interval of time approximately 541 million years ago in the Cambrian Period when practically all major animal phyla started appearing in the fossil record. It lasted for about 13 − 25 million years. This alone distances the evolution logic from reality.

In a review article by L. James Gibson about a book by Stephen Myers, titled "Cambrian Explosion and Darwin's Doubt."[120] He writes: *Many different types of marine animals appear abruptly in the fossil layers, with no clear evidence of evolutionary ancestors in the layers below them and no series of evolutionary links connecting the various types.*

The pattern of abrupt appearances in the Cambrian Explosion is a problem for Darwinian evolutionary theory for a number of reasons. Darwin's theory is based on the gradual accumulation of minor changes over long periods of time, eventually producing changes so large that new types of organisms could be recognized. If the theory is correct, the rocks should contain a long series of intermediate forms preceding the appearance of different types of animals. The fossil record does not record such a pattern, and Darwin acknowledged that.

---

[120] In a review of the book "Darwin's Doubt" by Stephen Myers. James Gibson wrote an online article highlighting the discussion of the Cambrian explosion within the book and summarises important points.

https://www.grisda.org/cambrian-explosion-and-darwins-doubt-1#:~:text=Another%20 problem%20for%20Darwinian%20theory%20is%20that%20all,by%20the%20late%20 evolutionary%20scientist%2C%20Stephen%20Jay%20Gould.

E. All the major types of animals were present in the Cambrian, yet they are represented by relatively few variations. This pattern was dubbed "disparity precedes diversity" by the late evolutionary scientist Stephen Jay Gould. This pattern is not merely different from what Darwin's theory predicts, but it is the exact opposite of the expected pattern. It should take many small changes to add up to a large difference in a new type of animal, but the small differences appear in the fossil record after the large differences. Moreover, this pattern is not limited to a few phyla but applies to the entire animal kingdom.[121]

F. For evolution to develop in an upward fashion, where species get better all the time, you probably need a driver behind it. If they were as random as evolutionists claim, then the design would have taken many dips rather than getting better all the time, which would be unrealistically lucky.

G. Our strongest and most powerful attributes of design as humans are intellectual power and speech, something which was bestowed on us only. While other species seem to excel in other design parameters, for example, vision, speed or flying capabilities. This points to a very precise and intentional selection, in which different species were given different qualities, rather than a string of evolution where every species gets better abilities by evolving from another of lesser value.

---

[121] Darwin's Doubt. The Explosive Origin of Animal Life and the Case for Intelligent Design. By Stephen C. Meyer June 3, 2014.

H. Darwin writes: *"Geology assuredly does not reveal any such finely graduated organic chain; and this, perhaps, is the most obvious and gravest objection which can be urged against my theory."*[122] The obvious attempt at explaining the species existence lies in the observation of form, without looking at function, structure or behaviour. As the theory of micro and macro genetic mutations was of course unknow, at this stage of history.

I. Vision: Perhaps vision summarises the dilemmas puzzling philosophers, scientists as well as anybody who tries to understand the Evolution Theory.

First is how one can explain a logical process by which species decided if they develop eyes, they will be more survivable; claiming it was just a chance is nonsensically lucky. The second is for vision to develop; not only we need to metamorphose an eye structure, but there has to be a simultaneous visual centre in the brain which has to be wired at the same time, so which one develops first, and was there species with one and not the other till the evolution decided to interfere and rectify the situation?

In addition, writing the code for its development in the DNA and including these steps in the embryonic development stages and later growth till its full maturation.

---

[122] He adds to the above quote: "The explanation lies, as I believe, in the extreme imperfection of the geological record. In the first place, it should always be borne in mind what sort of intermediate forms must, on my theory, have formerly existed. I have found it difficult, when looking at any two species, to avoid picturing to myself forms directly intermediate between them. But this is a wholly false view; we should always look for forms intermediate between each species and a common but unknown progenitor, and the progenitor will generally have differed in some respects from all its modified descendants." http://www.talkorigins.org/faqs/origin/chapter9.html

And if one can explain all the above, we face a bigger dilemma, where science informs us: lesser organisms have a much more complicated eye structure than advanced humans. The rule is; complicated eyes go with simple brains.[123]

All the above points to intention and direction, and is not what nature is known to be capable of.

The Genetic Argument

DNA was discovered in 1868 by a Swiss medical student named Johann Friedrich Miescher. However, it wasn't until 84 years later, in the early 1950s, when scientists recognized it as genetic material.[124]

A. Post Darwin's scientific discovery of the DNA would have cast doubt over his hypothesise. Still, by then, the movement had taken shape and there was too much to lose, so they used the logic that there is DNA in all species and similarity to humans ranges from 50 % in rates to 95% in some types of chimpanzees. The percentage of similarity varies incredibly in many reports as it seems the way DNA sequencing is compared can be very alterable. When DNA is compared by single protein-to-protein comparison, they give an impression of high similarity, but once you compare the entire genomes, and use more rigorous methods, the similarity drops.[125] But again, logically, this resemblance in

---

[123] Eye and Brain The Psychology of Seeing, by R. Gregory, 5th edition, Princeton University Press: page 25, ISBN: 0-691-04840-1.

[124] https://www.dna-structure.com/history.htm

[125] https://idthefuture.com/1713/

the building blocks does not necessarily prove commonality in origin but rather a similarity in design technique. The first needs way more evidence than is currently available.

The highest percentage of similarity acknowledged between both humans and chimpanzees have reported a genome size of roughly 3 billion nucleotides for each. Yet they differ by 30 million fixed nucleotides."[126]

B. Haldane's dilemma:[127] According to evolutionists, humans and chimpanzee species diverged from a common ancestor approximately 4.5 to 13 million years ago. Human generation interval is considered to be 30 years, while chimpanzees is around 20 years.

Suppose we divide the highest estimated number of years by the number of a generation, 13,000,000/30 = 500,000 generation intervals. Haldane claims that if we consider the 30 million fixed nucleotide differences, we know exist between humans and chimpanzees, and we allow the maximum time of the existence of chimpanzees, we have to account for 3333.3 evolutionary changes. This puts the rate of DNA change at 1.5 per year in each line of descent, or 30 DNA changes per chimpanzee generation interval. This is a "BIOLOGICAL IMPOSSIBILITY." Adding to this, every addition or change must be favourable and beneficial, eliminating the possibility of deleterious mutations, which

---

[126] The Origin of Genome Architecture by Michael Lynch, History and Philosophy of the Life Sciences. Vol. 30, No. 3/4 (2008), pp. 491-493.

[127] Haldane's dilemma, also known as "the waiting time problem," is a limit on the speed of beneficial evolution calculated by J. B. S. Haldane in 1957.

J. B. S. Haldane, "The cost of natural selection," Journal of Genetics 55 (1957): 511–24.

makes evolutionists unrealistically lucky, as this is not how things happen in the universe.

C. In 2008, two biologists, R. Durrett and D. Schmidt, conducted a study,[128] where they examined the regulatory sequence evolution in Drosophila (genus of flies misleadingly called fruit flies) and compared it to humans. They examined the waiting time for a pair of mutations, the first of which inactivates an existing transcription factor binding site and the second creates a new one, they concluded that for the Drosophila, a few million years is sufficient, but for humans with a much smaller effective population size, this type of change would take more than 100 million years. Thus, they questioned the mathematical limits to the Darwinian evolution.

Since humans and humanoid species existed only 200,000 to 350,000 years ago, according to the oldest fossil evidence discovered so far, this just doesn't add up in the favour of evolution.

D. Micro and Macro genetic changes: In 1937, the evolutionary biologist Theodosius Dobzhansky invented the term microevolution[129] i.e., little changes or adaptations of animals to their environment. But macroevolution is the significant change from one species to another. Darwin's description of little changes at a time contradicts the most common consensus that there has to be a major macro change to get species change.

---

[128] Waiting for Two Mutations: With Applications to Regulatory Sequence Evolution and the Limits of Darwinian Evolution, Genetics, 180:1501-1509 (2008).

[129] https://en.wikipedia.org/wiki/Evolution

E. The most compelling evidence to date comes from a recently published study in Nature magazine in January 2022, titled "Mutational bias reflects natural selection in Arabidopsis thaliana" by J. Monroe et al., from the Department of Molecular Biology, Max Planck Institute for Biology, Tübingen, Germany and the Department of Plant Sciences, University of California, USA.[130] They formulated their study by testing the assumption of the evolutionary theory that mutations occur randomly concerning their consequences. They did an extensive survey of de novo mutations in the plant Arabidopsis thaliana. In contrast to their expectations, they found mutations occur less often in functionally constrained regions of the genome. Also, mutation frequency is reduced by half inside gene bodies and two-thirds in essential genes. They then concluded that epigenome-associated mutation bias reduces the occurrence of mutations in Arabidopsis, challenging the prevailing paradigm that mutation is a directionless force in evolution.

It was essential to present the above study in a scientifically accurate way for the hardcore researcher. To explain this in layman's terms, it means the following: The genetic material is divided into 1% of coding DNA, and the rest is noncoding DNA.[131]

The coding section is known as an exon, which is the sequence used to manufacture proteins. Coding DNA sequences are separated by long regions of DNA called introns with no

---

[130] Monroe, J.G., et al. Mutation bias reflects natural selection in Arabidopsis thaliana. Nature 602, 101–105 (2022). https://doi.org/10.1038/s41586-021-04269-6

[131] https://medlineplus.gov/genetics/understanding/basics/noncodingdna/

apparent function or at least were thought of as such.

The 1%, which is the coding DNA, is then further subdivided into a small segment of essential genes and a major section of non-essential genes. This latest study confirms a significantly lower mutation probability in the section of the essential gene compared to other sections of the genes, which indicates a protective mechanism built-in which protects these essential genes from mutations. As these are the important genetic sections of the DNA, where mutation needs to take place to impact macro genetics, this puts the mutation probability further away from the truth. They also state this was done on one plant, and other species need to be examined.

I would predict we will never see the end of the Darwinian vs. Anti-Darwinian argument as long as one side claims the validity of a scientific theory pending further discoveries. Or claiming the lack of an alternative logic on one topic proves whatever logic is available.

This is precisely what you can label as false logic. Or logical fallacies, which we have alluded to earlier. Where one argues a proposition must be true because it has not been proven false or there is no evidence against it. For example, when one says: "No one has ever been able to prove extra-terrestrials do not exist, so they must be real." Remember the "Appeal to Ignorance" fallacy or "Argumentum ad ignorantiam" we mentioned earlier.

ARTIFICIAL VS. BIOLOGICAL INTELLIGENCE

An extra-terrestrial Mr. E.T., on his first visit to planet Earth with his family, were accompanied by a human guide, and upon touring a car manufacturing plant, they observed a unique phenomenon, or so they thought, where the factory is run almost exclusively by machines, and as they didn't see any operator on site, they ask where did humans find this marvel, thinking they were just lucky, finding this miracle before their planet did. Mr. E.T., not convinced humans could be so clever, quickly infers chance, explosion, pressure, heat and radiation could have kick started the process, as these are all he sees in the cosmos anyway; he argues further that a vast amount of minerals and iron are abundant in comets and exoplanets. The human guide laughed at his comments and began to explain that a tremendous amount of work is done on many levels for the factory to operate as it appears. The lack of an obvious operator moving things, is due to the use of advanced techniques humans have recently developed called "Artificial Intelligence." This work can only be done by the hard work of a humongous number of engineers, scientists, and technicians, but Mr. E.T. cannot see them while he's inside the factory.

MR E.T.'s family, on their second leg of the earthly visit, are taken to the "Evolution Museum," they are told, these museums are spread all over the world. They observe the models of evolutionary progress, the majority of which are artistic drawings rather than factual science. Upon seeing the microscopic shape and function of the human cells, MRS E.T. concludes this is an even bigger marvel than what she saw in the Car Plant. Insisting on meeting the engineers behind this intelligent design. Again, the family's opinion is ridiculed by the human guide and this time, they are told the Biological Intelligence (BI) they had just observed is all a result of explosions, pressure, heat and radiation.

MR E.T.'s daughter objects and tries to explain that all of the factors mentioned above are known to be detrimental to any biological cell growth rather than progenitors of life, according to their extra-terrestrial school curriculum anyway. Her comments are shrugged off, claiming one day, our great human AI scientists will be able to prove how humans and life were made without a designer. You can probably imagine how Mr. E.T.'s family will react to human logic.

The Biological Intelligence used to build atoms upon atoms in trillion configurations, leading to an individual cell, then massive repeated duplications which appear to the naked eye as self-driven, followed by the few hundred differentiations in shape and function of various cell lines, is certainly few million times more complex than the AI technology we have today, as we cannot comprehend its actual mechanism yet. Despite that, some of us conclude AI could not have made itself, while BI is self-made by chance.

IV

The Designer

·

From antiquity to the present time, there are numerous documented claims of Deity. This ranged from Posthumous deification (declaration after the person's death) to Involuntary deification or Self-deification. None of which were short of followers, but most were entrenched in human exploitation. While if humans have used thinking and logic along the millennia, lack of exploitation should have been the main revealing attribute of any decreed Deity. Probably, colorful imagery and mystical legends facilitated the deceit to allure common people.

The oddity is, less than a handful of Deity claims have declared any knowledge or ownership of the human or the universe's design. Weird!

In our endeavor to look for the truth, we need to discuss what is undiscussable in scientific literature. Look for the expected characteristics of the "Designer" of the current universe by using attributes of the design, which we have agreed upon in the previous sections.

Most religions have extensively described the characters of a God, but we agreed we are not talking about religion. And to be honest, the regular theme of "God is love" as a character tends to sound more like a campaign slogan. We need to agree we will not put characters on this list unless we have proof thereupon from the evidence of design.

I once read about prophesizing of the God existing within each one of us; this is another example of an unfounded, illogical claim or a sect looking for a new recruiting strategy. We don't want here to look for an ideal character of an "Imaginary Designer," but rather the factual characters that have been exercised or are likely to be exercised on the human race. After all, this is our future and the future of humanity.

Looking up references on the topic, I found an interview by Lee Strobel questioning the late philosopher and scholar Antony Flew on his conversion from atheism to deism.[132]

Although his fellow atheists accused him of being senile, his deduction of the characters of the Designer in the interview seems to stem only from the assessment of the universe and human design, and is precisely what we are looking for in this context. The list we are about to compile here is only a suggested list and by no means exclusive or comprehensive.

FACTS ABOUT THE DESIGNER

Let us summaries the facts upon which we will list the characters we can attribute to the Designer: The aim is to use facts on the ground we currently know about our universe and the history of humanity to construct an idea about the Designer, which will give us a clearer insight about the Designer's logic.

---

[132] Antony Flew on God and Atheism, Feb 2013 Lee Strobel interviews philosopher and scholar Antony Flew on his conversion from atheism to deism. Much of it has to do with intelligent design. Flew was considered one of the most influential and important thinkers for atheism during his time before his death. His conversion to God-belief has caused an uproar among atheists. They have done all they can to lessen the impact of his famous conversion by suggesting he's too old, senile and mentally deranged to understand logic and science anymore. Quote from the publisher of a video interview. https://www.youtube.com/watch?v=VHUtMEru4pQ&t=606s

We are comprehending a "Designer" who has originated and created matter, anti-matter, based on tiny particles (protons, neutrons, and even smaller) made from mass and energy, held by four physical forces, occupying space within nothingness and wrapped in time.

A Designer of something cannot be bound by his design. Thus, it would be a logically valid deduction; he cannot be bound by time or the four physical forces, controlled by them, or even made from them. A creator of human consciousness cannot be bound by this either, which means we humans can never grasp his nature. Thus, we can surely conclude what he isn't rather than what he is. This is different from realizing his work, his characters, or his presence.

CHARACTERS ATTRIBUTED TO THE DESIGNER

We can hypothesize the Designer has the following attributes:

Constant: We will use the steadiness and the stability this term implies and leave the continuous or the endless it might suggest. (As the latter will come with other attributes). Remember consistency is a synonym for logic. We can agree about the presence of mathematical logic in this universe and leave philosophical logic aside. Thus, the creator of this universe definitely has consistency in the design process and sustainability.

For example, no scientist has ever predicted the laws of physics have changed over time or are even likely to change. This consistency is taken for granted by humans, but what if this is not the case? In other words, if our universe was a result of a haphazard process of occurrence, and consistency would be lost, and one could see a constant change in the laws of physics. I could leave this scenario to the imagination of the reader to contemplate how life would be like then. Or suggest to a Hollywood screenwriter the idea that scientists might notice a gradual change in the speed of the electron and let him imagine the ensuing catastrophes. Yet, we take this consistency for granted.

He will be consistent in the future (our future) as he was consistent in our previous history and everything he has created so far. He will consistently provide for humans until they depart from this world. We can only abduct this kind of reasoning by looking back at the information we know from the past.

Is the design over and done with, or is it ongoing? Physical evidence shows humans are born every day, babies grow up every day, galaxies explode, and others are reborn constantly. The universe is expanding.

If you leave an automated factory unattended for some time, errors start to appear, and eventually, minor errors start to build up, leading to catastrophic destruction. Thus, we need the maintenance engineer to keep it going. We have already observed the development of errors at the levels of biological cells and organs. If we do not conclude his main character is a constant Designer and maintainer of the current universe, we will be further away from the truth.

There could be endless examples of this attribute, and I will leave it to the reader to create a possible list in his/her mind.

Eternal: Most definitions on this topic are religious theology, which I will not refer to. But, logically speaking, if we know now the universe was made of particles which exist in space and time (scientific fact), this makes time one dimension of creation and design. Thus, you can't create time and be bound by it but rather out of its limits. Putting this in a simple logical phrase would be "Eternal."

Omnipotence: Which is the quality of having unlimited power.

Omniscience: Which is the capacity to know everything. We presume here everything in the universe and the coming meta-universe.

Intelligent: Which is by default after having built an intelligent design.

Caring: By creating a suitable habitat for humankind. As well as animal kingdoms. As we can infer, "Planet Earth" is a purpose-built habitat. And if humans had been created on other planets or without sufficient resources, their life would have either been miserable or probably ended. It is appropriate here to question atheists why chance has not been able to create life on other planets. Their answer is always on earth, there is water. But honestly, have they not thought on a comparable planet, chance has not been able to create beings that survive on methane gas, which is abundant on other planets? Scientists have found a significant amount of methane on Mars, and it is believed Saturn's moon, Titan, is awash with it.[133] However, even this explanation or reason is not a valid argument any more since NASA announced in April 2015 they have found water in the universe,[134] Then, later, on the

---

[133] https://www.scientificamerican.com/article/methane-on-mars-titan/

[134] The Solar System and Beyond is Awash in Water, April 2015: The chemical elements in water, hydrogen and oxygen are some of the most abundant elements in the universe. Astronomers see the signature of water in giant molecular clouds between the stars, in disks of material that represent newborn planetary systems, and in the atmospheres of giant planets orbiting other stars.

https://www.nasa.gov/jpl/the-solar-system-and-beyond-is-awash-in-water

moon,[135] Thus, the lack of water logic is not relevant any more. It's the environment as a whole, which is a purpose-built habitat, as we can see on our planet, rather than some elements that support life. This sophisticated dwelling is very strong evidence for "Caring."

Guardian: The Earth, the galaxy, and the universe we live in have demonstrated an incredible amount of locked up energy and power; without a "Guardian," we could have been history. This probably explains why we have such a steady life, and all of us can drive our children to school in the morning. The most compelling guardianship attribute one can demonstrate is the atmospheric shield, which has no comparison on other planets (despite that all planets have atmosphere) where we are protected not only from meteors, meteorites, but also from harmful radiation and temperature fluctuations. When you play in your garden with your children, just contemplate that someone is holding a massive shield over your head so you can feel so great and comfortable. Using other direction of logic as saying the protection is what allowed life to grow on its own, is like believing if you design a wallet, paper money will grow inside with time.

Just: Certainly, the design process has demonstrated justice and fairness at the lower cross-sections, but is lacking at higher levels of the human being and the human race. The strongest justice system we can observe lies in our own bodies as we have demonstrated

---

[135] NASA released a news announcement in October 2020: NASA's SOFIA Discovers Water on Sunlit Surface of Moon. NASA's Stratospheric Observatory for Infrared Astronomy (SOFIA) has confirmed, for the first time, water on the sunlit surface of the Moon. This discovery indicates water may be distributed across the lunar surface and not limited to cold, shadowed places. https://www.nasa.gov/press-release/nasa-s-sofia-discovers-water-on-sunlit-surface-of-moon

earlier. We hypothesized that the unfairness we see in the mentioned cross-sections is ascribed to humans.

We cannot escape the notion that, within this design process, innocent people do get hurt. Thus, going back to the argumentative logic we used, which is if a Designer is capable of exercising this level of justice at multiple levels, he most certainly won't leave injustice in these two levels (the human and human race) unattended; we also brushed on the reason why corrective justice cannot happen in this universe, with the current human race scenario. And fulfilment of this justice must happen in a meta-universe in the future. Otherwise, the whole exercise of conducting an exam within the universe would be a waste.

Originator: This attribute of the Designer, I personally visualize in many situations; the first is when you look at a natural reserve with no human intervention or influence. Nature always looks so beautiful, something all of us have experienced. Ranging from the tropical rain forests with their covered canopy of greenery and running rivulets of water to bursting volcanos with their angry lava, creating new fertile islands in the middle of the ocean. The second you can hear in the narration of physicists and astrologists talking about the cosmos, with passion in their speech and a glow in their eyes. The most compelling and one which fills your consciousness with adoration is when you put your finger in the hands of your own new born baby, when she is few hours old and feel her tightening her grip. Contemplating how this tiny delicate hand has all the nerves, veins, arteries, bones and muscles to do so. I am sure many people have felt the same.

Mercy: It is very difficult to quantify mercy and define it scientifically when discussed in the context of the universe and human design. However, I can demonstrate some evidence of mercy using logical reasoning:

Ongoing Freedom: I mean here freedom given so far to the human race to claim as they wish, who has designed them; after all, we have few thousand proclaimed Gods, so we have to presume definitely, there are many erroneous conclusions. Yet these humans continue to live and enjoy the freedom and the care all other humans get. This is obvious in the current universe, at least. Lack of mercy would have led to a different pathway for the human race. However, will this continue in the following meta-universe?

Human Need Vs Availability: If we contemplate the range of things humans need to survive on this planet, one can see an ingenious way of resource distribution.

If we attempt to create a hypothetical list commencing with the absolute need followed by dire essentials, then to important elements, and finally ending with luxury needs. This will look as shown in Figure 2.

One can go further, but the idea is clear. You will notice "Availability" is directly proportional to the "Need." For example, the first need is "Air," which is of absolute urgency and thus is readily available independent from any human intervention or effort on one's part.

Hierarchy of Human Needs:

- Air.
- Water.
- Food.
- Shelter.
- Transport.
- Communication.
- Sports.
- Entertainment.

Figure 2: List of human needs.

Second is "Water," which is widely available but needs some effort to obtain and is followed by other needs which require harder and harder work to acquire and revolve around humans depending on and needing each other, for if this was not the case, humans wouldn't have lived in groups or cooperated together. The mercy of the design gave us what we essentially needed and allowed us to cooperate together to obtain more sustenance and comfort in life.

Human Disease and Mercy: Contrary to the atheistic claim, I clearly see human disease as a merciful attribute rather than a design fault or a wrath upon us. This might need some explanation as it might contradict the obvious, which is pain and suffering.

- *Self-limiting:* Many diseases affecting humans daily are classified as self-limiting, and it is precisely so because of the mercy of the design process that made us. How the human cell is a dynamic living unit which is constantly monitoring and adjusting its processes, working to restore itself according to the DNA code it was created with, and to maintain balance within the body. Cells can heal themselves, as well as make new ones to replace those permanently damaged or destroyed. The mechanism of how cells are triggered to start the repair process and end it, is yet unknown to humans. When the inhibition mechanism of cell replication stops working, cell growth becomes unstoppable, and the human body develops tumours and pathological growths.

- *Human Induced diseases:* A great number of serious diseases, like malignancies and errors in metabolism, have proven to be the result of human intervention, like nuclear explosions and pollution. You only need to look at the rise in the incidence of malignant disease and congenital malformations in countries who were in a war scenario to be more convinced. The irony is the academic literature is full of studies to show the rise in malignancy in war veterans,[136] [137] who were dropping the bombs, with only news reporting[138] and doctor's observations from the countries on the receiving end.

---

[136] McDiarmid MA, et al. Surveillance of Depleted Uranium-exposed Gulf War Veterans: More Evidence for Bone Effects. Health Phys. 2021 Jun 1;120(6):671-682. https://pubmed.ncbi.nlm.nih.gov/33867437/

[137] Bjørklund G, et al. Depleted uranium and Gulf War Illness: Updates and comments on possible mechanisms behind the syndrome. Environ Res. 2020 Feb;181:108927. https://pubmed.ncbi.nlm.nih.gov/31796256/

[138] Aitken M. Gulf War leaves a legacy of cancer. BMJ. 1999 Aug 14;319(7207):401.

The conference (about the cancer epidemic in Iraq and its possible link to the Allied use of depleted uranium weapons) was chaired by Labour backbench MP George Galloway. The UK Ministry of Defence has declined to comment to the BMJ on the health implications of exposure to depleted uranium during the Gulf War. An Iraqi oncology specialist, Dr. Mona Kammas, presented a report compiled by Iraq's Committee for Pollution Impact by Aggressive Bombing. Rates of cancer and congenital anomalies had almost doubled since the war, the report said.

https://www.ncbi.nlm.nih.gov/pmc/articles/PMC1127036/

The Centre for Disease Control in the USA (CDC) have published on their website: Smoking can cause cancer and then block your body from fighting it.[139]

They state poisons in cigarette smoke can weaken the body's immune system, making it harder to kill cancer cells. Poisons in tobacco smoke can damage or change a cell's DNA.

They add in the United States, more than 7,300 non-smokers die each year from lung cancer caused by second-hand smoke. This is the combination of smoke from the burning end of a cigarette and the smoke "breathed out" by someone smoking. Smoking can cause cancer almost anywhere in your body, including the blood (acute myeloid leukaemia), bladder, cervix, colon, rectum, oesophagus, kidney, larynx, liver, lungs, trachea, bronchus, mouth, throat, pancreas and stomach.

Despite, this information is readily available and well known to all governments and individual humans, there is no action taken to stop the drain in human health or life. As it is claimed to be against human's right to choose. Thus, it seems humans want the design process to allow them to destroy anything they want and not suffer the consequences.

I would like to suggest the reader does a search about the introduction of the leaded gasoline by a man called "Thomas Midgley," who was labelled "The Most Harmful Inventor in History."[140] He discovered that ethanol (the same ethyl alcohol found in wines and spirits), can

---

[139] https://www.cdc.gov/tobacco/campaign/tips/diseases/cancer.html

[140] https://www.bbvaopenmind.com/en/science/research/thomas-midgley-harmful-inventor- history/

stop engine knocking, and in 1920, he filed a patent application for a mixture of alcohol and gasoline as an anti-knock fuel. When this was turned down, as it was readily available and people could distil their own ethanol at home, it didn't please the car industry. Thus, he was sent again to discover something else. In 1921, Midgley had found the solution: tetraethyl lead (TEL). A very cheap lead containing chemical. The deadly effect was obvious very early afterwards, yet it wasn't till the year 1980 when awareness started to phase this chemical out. The last stock was phased-out in Algeria in the year 2021. Yet it is still allowed for use in some aircrafts, racing cars, farm equipment, and marine engines. Within the same year of his (TEL) discovery, he discovered Chlorofluorocarbons (CFC)[141] for fridges, which destroyed the ozone layer.

Huge amounts of lead remain trapped in the soil, air, water and in our bodies. According to a 1992 article in The New England Journal of Medicine,[142] The average levels of lead in the bones of modern people are 625 times higher than those of the pre-Columbian inhabitants of North America.

According to the Centre for Disease Control in the USA, people with prolonged exposure to lead are at risk for high blood pressure, heart disease, kidney disease, and reduced fertility. Does this ring a bell?

The Department of Health and Human Services (DHHS) USA, the Environmental Protection Agency (EPA) USA and the International

---

[141] Chlorofluorocarbons (CFCs) are fully or partly halogenated hydrocarbons that contain carbon (C), hydrogen (H), chlorine (Cl), and fluorine (F), produced as volatile derivatives of methane, ethane, and propane. They are also commonly known by the DuPont brand name Freon.

[142] Lead levels in preindustrial humans. By A. Flegal, D. Smith. N Engl J Med. 1992 May 7;326(19):1293-4.

SCIENCE AND FAITH

Agency for Research on Cancer (IARC) have determined lead is probably cancer-causing in humans.[143]

As a mental exercise, try and look for diseases you come across in life, and search the incidence of this group of diseases in the 21st century compared to 200 years earlier. And look for possible connections between modern civilization and disease. I would suggest to start with anxiety and depression and I am sure anyone who does a search on the topic will be shocked.

It is agreed, older times had many other endemic diseases human civilization managed to combat, but in their endeavor to gain more power and richness, they have caused a multitude of new ones.

We can deductively conclude its freedom of choice and thus human action, i.e., human unfairness that produces a significant number of diseases. The survival of humans from these diseases and this perpetuated unfairness is in big part due to the merciful nature of the design process and the "Designer."

We still have to agree that human disease happens, even if not as a result of iatrogenic intervention. But let us look at human disease from another perspective.

- *Self-Reflection:* A significant advantage of having human disease, besides the fact people like me get to make a living, is it encourages human beings to look and study their bodies and marvel at their design and performance. If we were as accurate and faultless as an iron atom, we probably would have been less inclined to explore our form and function.

[143] Health Problems Caused by Lead.
https://www.cdc.gov/niosh/topics/lead/health.html#:~:text=Exposure%20to%20high%20levels%20of,a%20developing%20baby's%20nervous%20system.

Maybe then, the human disease was created to encourage us to appreciate and examine our design and learn how we were made, and this will lead us to our Designer.

- *Empathy:* Human disease is a significant reason for humans to develop empathy towards each other, either socially or even professionally, in the form of health-related industry. Empathy-related responding, including caring and sympathetic concerns, is thought to motivate prosocial behaviour, inhibit aggression and pave the way to moral reasoning.[144]

There is evidence prosocial behaviours such as "altruistic helping" (Altruism is acting to help someone else at some cost to oneself) emerge early in childhood. Infants as young as 12 months of age begin to comfort victims of distress, and 14 to 18-months-old children display spontaneous, unrewarded helping behaviours.[145]

As disease is a big perpetrator of distress in humans, we can infer its existence will help develop empathy in humans from their early years of life.

- *Examination Hall Concept:* On a final note, it is appropriate to link the presence of human disease with the examination hall hypothesis we discussed earlier. It certainly does make the exam a bit more challenging.

---

[144] Eisenberg N, Eggum ND. Empathic responding: sympathy and personal distress. The Social Neuroscience of Empathy. Cambridge: MIT Press; 2009. pp. 71–83.

[145] Warneken F, Tomasello M. The roots of human altruism. Br J Psychol. 2009;100:455–471.

SCIENCE AND FAITH

Human Death and Mercy: We all have a finite amount of time to spend on this planet and in this universe. Our demise is encoded in our DNA, this allows aging to creep on us gradually but persistently, and at the end, we all have to depart. The time span we have is widely variable among individual humans and does not depend on any background, race, power or wealth. To put it in a logical sentence: some of us are born dead, and others get to be centenarians.

There is no fault or aberration in the death mechanism; at the highest estimates, rare villagers were found to live till they were 120 years old. Why do some genetic mutations have never, by chance or error, allowed somebody to escape death? In over 80,000 years of history in human existence, this has never happened. This precision adds to the logic of a predetermined process rather than an un-directional creation. Now, to the logical mercy of death.

Imagine if Nero the Roman emperor, or Adolf Hitler the Nazi Leader, or Ivan the Terrible the Russian Tsar, got to live forever just because they were rich or powerful. What would be for the rest of us? Probably, most humans, if they had a chance to escape death, would change into dangerous creatures, threatening the future of the human race. (No inferences to prove that, but just spend a few minutes contemplating the idea, and it will be more apparent in your mind). The closest guess I can come up with is humans will be forming militias to fight newborn babies as they will be termed economic parasites who want to suck our wealth. This thought alone makes me glad to be among the mortals.

If you find it hard to believe this could be possible, try to search for the term "de-population," and don't refer to the conspiracy theory proponents that will dominate the search results but look at the study published in the Journal BioScience,[146] in 2020, stating: *"World population is increasing by 80 million people per year, and must be stabilized—and, ideally, gradually reduced—within a framework that ensures social integrity."* These claims can be refuted by other scientists as not of any concern but are used by a few extremists, lamenting the idea the earth is overpopulated and we must scale down our numbers. I expect in the future, we will be hearing more about this.

Thus, we can deduct mortality is essential for the human race's survival. The next topic to discuss is how merciful is the death mechanism applied to the human race.

By imagining the alternative scenarios of death falling upon humans, how will that be applied in any different way?

Scenario A: We could all be allocated a definitive equal time on this earth, and then we die, let's say hypothetically, at the age of 70 years, with no aberrations. The idea sounds fair, but let us imagine the consequences:

- Humans will stop fearing death during the majority of their lifetime, and this will result in a riskier lifestyle to themselves and to other fellow humans.

- What would happen to those fellow humans who develop disease or face accidents. Will they endure a lot of suffering till they reach the pre-set date? We know from life experience that passing away of chronically ill patients is looked upon as an attribute of mercy. One can only imagine these humans will be dying to die.

---

[146] World Scientists' Warning of a Climate Emergency. By W. Ripple et al. Journal BioScience, Volume 70, Issue 1, January 2020, Pages 8–12, https://doi.org/10.1093/biosci/biz088

SCIENCE AND FAITH

- If we contemplate this idea, one can imagine how humans will attempt to cheat the system, as some smart students do sometimes: play most of the academic year, then sit on the books a few weeks before the exam. But humans themselves have created a mechanism to avoid this behaviour by incorporating school year work into the marking system or doing regular emergency exams. Thus, I guess not knowing when we depart from this world will keep humans working hard and be ready for their assessment. We can use the variable and unknown departure time from this universe as evidence on the validity of the examination hall hypothesis.

Scenario B: We could suddenly die at any stage of our life; although this still happens, it is not the norm, and most people have preceding signs and symptoms, which logically leads to death. Lack thereof would lead us humans to be unable to live our lives, or we would be living with constant fear, as you would expect life to end at any second without any preliminaries. We haven't asked people who lost their life suddenly, what they would have preferred? But being able to predict ahead is a great mercy, despite being of limited time on some occasions.

Scenario C: We could get a sign in our body, like when plants ripen to indicate we are about to go. Imagine how one's life would be like from the time you get this sign till you die. We can probably ask people on death row, as I am sure the feeling will be comparable.

One can argue we still get gentle warnings like graying of our hair, wrinkling of the skin, decline in power and increase in the number of candles on our birthday cakes. All of the above are never a logical reason for death; however, disease is.

To summarize my view on linking human death with mercy, variable forms of diseases, which might or might not end up with our death, with multifarious escapes but ultimate demise, are probably the best merciful ways to depart from this universe to the meta-universe.

Further discussion on Human Death will come in chapter V when we discuss human fear of the unknown.

This brings us to the discussion of the design flaws and aspects of human weakness.

V

The Design Flaws... Or Is It?

HUMAN DESIGN WEAKNESSES

The previous discussion has highlighted inherent weaknesses in the design, and it is quite evident these are built-in attributes of design and not a design fault. One can question evolutionists: Why hasn't inherent weakness manifested evolutionary improvement?

We list four very important manifestations of weakness we meet in our lives. Most if not all are common reasons mentioned by atheists to explain their believe system, inferring there is no "Controller" of the current universe.

SUSTENANCE AND DEPENDABILITY

In the minds of a significant group of humans the question: Do we need the Designer anymore? After all, we have everything. Could we be independent of him? We need to get a valid logical deduction.

We were designed with a very small fuel tank; imagine a machine that can perform so little and then needs a constant supply of fuel and rest. Hunger and fatigue factors ensure our powers are limited in force and duration. This becomes more obvious when we compare our design to some animals, who can demonstrate the ability to live for days on a single meal and perform even better.

Sustenance Obsession: Past human history of famine, our constant reminder of feeling hungry, and perhaps greed, made humans and the human race obsessed with sustenance. So, we not only work hard to secure our daily living, but we also want to secure the future and decided to create reserves of sustenance, which we might or might not need. We went even further, using sustenance as metrics of our daily life success, jobs, and careers. Worse is how we formulate our opinion about fellow humans who have extra sustenance; we start looking at them with respect and envy. Thus, we conclude that, as an integral part of our design, we constantly need sustenance. But why?

Philosophically, the constant feeling of hunger makes us appreciate satiety and fullness, the feeling of weakness makes us appreciate strength, and the feeling of need makes us appreciate affluence. Thus, we can deduce we were intentionally designed to constantly feel deficient, and the obvious explanation is it forces us to appreciate the opposite. Therefore, it serves as a reminder that we do need sustenance, and we are dependent on the design process, thus by default on the "Designer."

FAMINE AND HUNGER

In such an intricate design process, it is astonishing to see famine and starvation, one of the most devastating catastrophes experienced by humans. Why do we see famine and hunger despite this intricate and intelligent design? This question has irritated my brain for many years till I did a simple search about "Causes of Famine."

According to the Encyclopedia Britannica, famine is defined as *Severe and prolonged hunger in a substantial proportion of the population of a region or country, resulting in widespread and acute malnutrition and death by starvation and disease.*[147] The strange fact is famine usually lasts for a limited time, ranging from a few months to a few years. And never indefinitely.

Until the 1980s, the underlying causes of most famines were poorly understood, and strangely enough, previously held beliefs it is an act of God have proven not to be true.

Three Major Causes of Famine

Warfare: According to the same source, warfare is the most common cause of famine. Not only because war can directly damage crops or prevent farmers from planting and harvesting, but also the deliberate destruction of crops and food supplies became a common tactic of war since the 19th century. Adding to that war often cause mass migrations as people flee the fighting. Displaced populations in refugee camps may not have access to their usual food sources, making them dependent on external aid, which might be insufficient.

Entitlement Failure: In 1982, Amartya Sen challenged the prevailing assumptions that total Food-Availability Decline is the central cause of all famines. He argued that the more accurate cause is "Entitlement Failure,"[148] which can occur even when there is no decline in aggregate food production. Examples of entitlement failure are stock market crashes, drop in commodity prices leading to income reduction, and bad political moves by governments facing elections or dissent. In most of these situations, there could be only a minor decline of food production or even normal production, but

---

[147] https://www.britannica.com/science/famine

[148] https://www.britannica.com/science/famine/Entitlement-failure

failure in distribution of resources or poor decision-making is what leads to famine.

Drought: Most scientific references agree that drought can be a direct driver of famine, however they indicate that other underlying vulnerabilities like poverty, lack of infrastructure, political instability, and poor governance can amplify its impact.

I can understand drought, a thousand years ago, would have been a problem, but humans in those days understood they could move to a better place as weather systems changed. But with the advancement of knowledge, humans started creating blockades against the movement of fellow humans and called them economic migrants; this led to the drought-stricken areas being less likely or able to relocate, creating disaster-stricken spots on earth.

In summary, most (if not all) causes of famine are either caused by humans or indirectly made worse by their decisions. The unusual fact here is famine lasts for a short time. Thus, it is fair to conclude that, despite human effort, populations survive their own doings.

HUMAN DISEASE

After writing the philosophical deliberations of why we do meat disease events in our lives, within the design flaws list here, I opted to move it to the signs of mercy section in chapter IV, as I strongly believe we humans do not grasp enough the intricacy of our design and how much self-correction is built in our systems, enough to make it more of a merciful attribute than a wrath. But the reader should mentally include this topic within the design flaws ponderings as well.

HUMAN FEAR

Fear is an affliction perturbing all humans and is one of six basic emotions.[149] If we contemplate our design's weakest spot, fear comes as the most widespread commonality among all mortals and is emphatically an integral part of our design. It dominates the human being and the human race from the moment they are born till they digress out of this universe. Individually, no one human can escape fear, even the ones we describe as fearless. It is a major underlying reason for aggression, thus unfairness, we discussed earlier. In this context, we will be talking about the functional concept of fear and will not include pathological fear, i.e., anxiety disorders, as we are interested in how normal humans are designed.

We are born with only two types of fear, "fear of falling" and "fear of loud noises." Later on, we develop other types of fears. Reiss's expectancy theory[150] states all fears arise from three basic types and are termed "Fundamental Fears": Fear of anxiety, Fear of negative evaluation, and Fear of injury/illness.

Functions of Fear

Why is fear incorporated in our Design? We agreed that for us to realize a feeling, we need to be wired to process this feeling.

Protective function: Humans rationalized the primary function of

---

[149] A widely accepted theory of basic emotions and their expressions, developed by Paul Ekman, suggests we have six basic emotions. They include sadness, happiness, fear, anger, surprise and disgust.

https://www.paulekman.com/universal-emotions/#:~:text=What%20are%20the%20six%20basic,seventh%20emotion%2C%20which%20is%20contempt.

[150] Reiss S, McNally RJ. (1985) Expectancy model of fear. In S Reiss and RR Bootzin (Eds). Theoretical issues in behaviour therapy (107-121). San Diego: Academic Press.

fear as a compelling mobilizer for an individual to deal quickly with important encounters. It alerts us and puts our body in an autoprotective mood, with the automatic trigger of fight and flight response we discussed earlier. Just imagine a fearless human being; he could be a very dangerous machine to human civilization. Humans would not have survived without fear.

We are pushed hard to ask for help: Another imperceptible logic for the cause of fear can be fathomed; when humans fail to find needed comfort from any attainable source, this potentially channels them to look for help from other humans or ultimately from their "Designer."

The literature is full of studies looking at religious coping mechanisms in patients suffering from anxiety and extreme crisis.[151] [152] In an original paper[153] About the topic, Niels Hvidt, a Professor at the University of Southern Denmark, writes: *"The majority of basic religiosity measures are associated with positive mental and bodily health outcomes,"* he also adds: *"Nothing seems to propel, activate and intensify religious seeking as much as mountains of crisis and disease."*

We have to conquer not all humans respond the same way when faced with difficulties, but it seems the design process allowed some of them to find relief in spirituality, and several improved as a result, according to scientific research.

---

[151] Religious Coping Methods as Predictors of Psychological, Physical and Spiritual Outcomes among Medically Ill Elderly Patients: A Two-year Longitudinal Study. By K. Pargament, et al. Journal of Health Psychology Volume 9, Issue 6, July 2016, https://doi.org/10.1177/1359105304045

[152] God Help Me: Advances in the Psychology of Religion and Coping. K. Pargament Archive for the Psychology of Religion 2002 24:1, 48-63

[153] Faith Moves Mountains-Mountains Move Faith: Two Opposite Epidemiological Forces in Research on Religion and Health, Journal of Religion and Health 56(1) February 2017. DOI: 10.1007/s10943-016-0300-1

In conclusion, Maybe by incorporating fear into our consciousness, we are pushed hard to ask for help.

Fear of the Unknown:

In my attempt to link fear to human design, I need to bring to the reader's attention another important type of fear: "Fear of the Unknown." This is defined as: "Fear caused by perceived lack of information or intolerance of uncertainty." A baby as young as four months old cries when she is held by an unknown person or is placed in an unfamiliar environment; she will not fear fire yet, as this will be learned with experience.

Lovecraft in 1927 said:[154] *"The oldest and strongest emotion of mankind is fear, and the oldest and strongest kind of fear is fear of the unknown."* Nicholas Carleton wrote a paper[155] in the Journal of Anxiety Disorders, titled: *"Fear of the unknown: One fear to rule them all?"* He utilized eight criteria to hypothesize this kind of fear as a fundamental type of fear, rather than an ordinary fear, to be included to the three types mentioned earlier.

Our Consciousness perception of the "Unknown" instinctively refers unknown issues to the brain's neural cognitive network, which tries to look for missing details using intuition skills; if only some missing data are encountered, it tries to fill the gap spontaneously or recalls previous events from memory to look for similarities or inferences which might fill absent explanations. If none of the above exists, then fear sets in.

---

[154] Lovecraft encyclopaedia. 2001. S.T. Joshi, D.E. Schultz. H.P. Greenwood Press, Westport, CT (2001).

[155] Fear of the unknown: One fear to rule them all? Journal of Anxiety Disorders. By N. Carleton. Volume 41, June 2016, Pages 5-21.
https://www.sciencedirect.com/science/article/pii/S0887618516300469

Our brain response to the unknown is elegantly described by Le Cunff on her website "Ness Labs."[156] She writes: *"Our brain is wired to reduce uncertainty. The unknown is synonymous with threats that pose risks to our survival. The more we know, the more we can make accurate predictions and shape our future."*

Congruousness of Fear of the Unknown with Design:

In attempting to discern the Designer's Logic from incorporating Fear of the Unknown in our consciousness, we can identify a conformity by looking at what was designed to be ultimately unknow for us.

Attempting to ascertain the foremost issues humans perceive as unknown and thus fearful during their finite presence on earth would be:

-   Fear of the future (what happens tomorrow).
-   Fear of death.

It logically follows we are meant to fear these unknown parameters.

What could be the Designer's intent from guiding us to fear from these events?

Why Do We Have to Fear What Happens Tomorrow?

In this matter, we are not totally blind, as we have formulas and equations which can predict many things in our lives, yet we still get

---

[156] The Uncertain Mind: How the Brain Handles the Unknown. By Anne-Laure Le Cunff, founder of Ness Labs and a PhD researcher investigating the Neuroscience of Education at King's College London.
https://nesslabs.com/uncertain-mind#:~:text=Our%20brain%20is%20wired%20to,essential%20gaps%20in%20our%20knowledge.

many surprises. This knowledge is dependent on a great effort and group hard work to attain. Logically, we need to plan our life and not just live as if there are always bananas on the tree. Or maybe we are hard-pressed to ask for help from fellow humans; thus, we are obliged to cooperate, which helps communities achieve better efficiency and reliability, as we have discussed earlier. This serves in strengthening the human race as a whole.

If we knew and could accurately predict the future, life would have been very boring or probably would have favored those who knew more than those who don't. Look at the stock market as an example.

Similar to the function of general fear, this will hard press humans or at least some of them to seek religiosity, thus communication with their designer.

Why do we have to fear Death?

There are two types of fear concerns to ponder here: one is fear of the timing (Death anxiety or Thanatophobia) and the other is fear of what follows (Apeirophobia).

Any realization of meta-universe life seems to be shrouded in the most rigid of boundaries humanity has ever encountered. It is so unknown to the extent some humans are more comfortable in conceptually subtracting this possibility from their daily life planning, while others have mentally classified it as mystical territory and then used their divergent thinking skills in imagining what it could be like, even more, some have started claiming it doesn't exist. Obviously, the design process has not paid much attention to the human deliberations and will not enable us to comprehend this part of the future; thus, the matter will continue being unknown till we cross that Rubicon.

Although the occurrence of death can sometimes be predicted, statistically or by health evaluation, like an individual who has chronic heart disease or many other types of chronic illnesses, or by virtue of age, an octogenarian has a more probable chance of worrying about meeting death than a teenager, but for the majority of humans, this is obscure information. We need to dig a bit deeper in the concept of "Fear of Death," which seems to be a common feature among all humans, to the extent, if this is lacking in an individual, it would be considered an abnormality rather than a variation of design.

When I meet people who claim they are not afraid to die, I get the impression what they mean is they are unlikely to meet death. All scientific studies concerning death anxiety relies on people's feedback of what they feel, but quit often, human feelings are shrouded in denial and are not real. The probable answer would be more accurate a few moments before this event happens, when one is certain he is just about to cross over.

Now to the function of "Fear from death"

- *Protective Function:* Similar to the logic of the "Fear" function we discussed earlier. There is a major protective role for fear of death. Otherwise, humans would just walk into fire like zombies and wouldn't care.

- *Convincing humans to seek the Designer's help:* A study[157]

---

[157] G. Morris, et al. March 2009. Are personality, well-being and death anxiety related to religious affiliation? Mental Health, Religion & Culture. 12 (2): 115–120. doi:10.1080/13674670802351856

in 2009, about death anxiety showed one religious group scored lower values for death anxiety than non-religious individuals, which supports the idea people pursue religion to avoid anxiety about death by finding comfort in the ideas about the afterlife and immortality. However, when this was tested in another group from an alternate religion, the results were not the same. Thus, it seems human perception of death anxiety is very variable and is perceived differently in each religious group. However, for some of them, religiosity seems to help in surmounting its effect.

- *Making Good Use of Available Time:* Le Cunff writes:[158] "Death anxiety is a fear of running out of time", which could be a principle design reason to direct us in making good use of our available time.

- *Searching for meaning:* The fear of death can drive some to seek a deeper meaning of life, whether through religion, spirituality, or philosophy. We are thus compelled to search.

- *Reminder:* For the question about the obscurity of "Death." If we link the examination hall hypothesis, together with the lack of fairness in this world we have proved at many levels, to the achievement of fairness prediction we have discussed earlier, this leads one to deduce our examination period (life) will end suddenly and our marking (meta-universe) will come as a surprise.

---

[158] Time anxiety: Is it too late? By Anne-Laure Le Cunff.
https://nesslabs.com/time-anxiety#:~:text=While%20death%20anxiety%20is%20the%20fear%20of%20running,your%20time%20in%20the%20most%20meaningful%20way%20possible.

Thus, we are supposed to be ready and not just study a few days before the exam. The notion is the design process has been setup up to make us fear what is coming; and this is the only information we can assimilate. Humans who care have been busy taking time to prepare and those who don't are only reminded by instilling fear in their consciousness.

Consequently, the most plausible explanation for it to be, ultimately and irrevocably unknown to us and in triggering our fear response, is a kind of alarm. WE ARE BEING WARNED!

Humans Reaction to Death Anxiety (Thanatophobia).

Some humans accept religious narration of what happens after death as a logical process that explains this unknown parameter; thus, to them, this is not unknown anymore; it changes into a "known unknown" (religious). For others, they attempt to mentally subtract the fear from this unknown and claim they don't care, creating an "unknown unknown," (agnostics) and they are content with the way things are. Further on, some try and convince their consciousness there is nothing, and as probably there is no need to fear nothing, this makes them more comfortable; they are then creating an "unknown known" (atheists).

I thus claim that individual human believe systems about the afterlife, thus religion, is deeply influenced by their perception of "fear of the unknown" and, more specifically, their "fear of death" and unambiguously fear from what follows.

To prove my point, I would like to refer the reader to a meta-analysis study[159] published in 2018 by Dr Jonathan Jong et al., who led a team in reviewing 100 studies in the literature about death anxiety among various believe groups, and they report very low anxiety levels in the very religious (in my previous explanation as they have reduced the unknown to be known) and atheists (as their consciousness perceives it as nothing so again is a known parameter thus no need to worry any more). They report higher death anxiety in the uncertain individuals who, in my previous narration, have kept things as (unknow unknown).

Where is the error in the human perception?

Could there be a cognitive bias or logical error in the above three intellections of apeirophobia? The answer is almost certainly yes, as the three avenues of belief are totally contradictory, i.e., there could be no both occurrences of something and nothing. What can we deductively elucidate?

In the known unknown (religious), many humans utilize abstract thinking, trying to connect information together to come up with an explanation. Many falls into the casual uncertainty bias[160] we discussed earlier, resulting in a multitude of faiths, many of them extreme, even claiming a human or a mystical figure as a deity. The coherence is; there should be no belief without evidence, and the latter must follow logic.

---

[159] Jonathan Jong et al. The religious correlates of death anxiety: a systematic review and meta-analysis, Religion, Brain & Behaviour, (2018) 8:1, 4-20, DOI: 10.1080/2153599X.2016.1238844.

[160] Casual uncertainty: As people seek to understand events and how everything relates to everything else, it can sometimes cause people to seek out pat-terns, themes, and relationships that may not exist. (Footnote 100)

The agnostics in the second group (unknown unknown) are those who decided they are not interested to find out and are content to keep the unknown as unknown. We can't oblige them to share our inquisitiveness, but I can refer them to the next section of why should we care and what happens if we don't.

In the third group, who decided the unknown is, in fact, known and is nothing, one can think of a similar parable, when humans were faced with many unknowns, for example, when scientists started studying the universe to find it empty and for a long time believed it is filled with nothing, now as they gained more knowledge they declare *"there is no such thing as nothingness in the universe".*[161] Similarly, when they studied the atom, believing the nucleus is surrounded by orbiting electrons with the majority of the space in-between being empty with nothing, now they believe the atom is full of particles and antiparticles and there is no nothing. Then, when studying the human genetic code, for many years scientists believed only 2% of it was useful and the rest they labeled as "junk" or "garbage" with no benefit. Now, after the advancement of their knowledge, they realized it has a function. There are many examples of humans being hasty in expressing opinions before acquiring knowledge.

Cavalierly believing there is nothing due to a lack of knowledge of the after-death scenario makes one infer they could have fallen into the same logical misconception.

---

[161] Quantum nothingness might have birthed the universe. There is no such thing as a void in the universe. M. Gleiser is a professor of natural philosophy, physics, and astronomy at Dartmouth College.
https://bigthink.com/13-8/quantum-nothingness-birth-universe/

VI

Challenging the human BRAIN

WHY SHOULD WE CARE?

Can we afford to be agnostic? After all, life still carries on. Why do we have to look for the Designer or a Message from the Designer? Why do we have to believe? We have previously alluded to the topic, but here we need to articulate the consequesnces of nonchalance inclination.

Current evidence in the universe demonstrates, both those who believe and those who don't live a comparably normal life. In logical terms, there is no statistical evidence showing those who believe and those who don't are any different in the quality of life, available resources, richness, or even life expectancy.

Research in this topic is a bit cluttered, with many papers claiming higher self-satisfaction in believers; some of these publications were done by researchers from the same religious background,[162] Others looked at multiple faiths.[163] On the other side, leading atheists have published papers claiming religious belief is a result of self-deception. Juha Räikkä, from the University of Turku, Finland, argued against this view.[164] His reason is based on the conceptual

---

[162] Relationship between the Religious Attitude, Self-Efficacy, and Life Satisfaction in High school Teachers of Mahshahr City. By M. Bigdeloo and Z Bozorgi. August 2016, International Education Studies 9(9):58.

[163] Koenig HG. Religion, spirituality, and health: the research and clinical implications. ISRN Psychiatry. 2012 Dec 16;2012:278730.

[164] Räikkä, J. (2014). Self-Deception and Religious Beliefs. In: Social Justice in Practice. Studies in Applied Philosophy, Epistemology and Rational Ethics, vol 14. Springer, Cham. https://doi.org/10.1007/978-3-319-04633-4_12

point of self-deception, mainly that self-deception must involve false belief. Which most believers in all sects of religions will not categorize themselves in.

In conclusion, academia and research have not been able to tell us whether we should or should not care about believing. And it is up to every individual human to pose this question to himself and answer the way he likes.

To demonstrate to the reader why it is imperative we find the truth, we will look at two issues: one is political, or so it sounds, and the other is hypothetical.

Political: All Emperors in the history of humanity have chosen a religion which they have used to their advantage. The primary aim was to gather funds and protect their realm by asserting that they were chosen by God. A secondary, less evident goal was to unify their territories and prevent conflicts among their subjects. What better way to do so than creating collective consciousness through faith. On some occasions, when they had political conflicts with their next-door neighbors who shared the same ideology, they splintered the faith and created a new sect overnight to support their quarrel. After all, how else could they persuade their soldiers to fight neighbors who held the same beliefs?

These Emperors knew when belief grips society as a whole and individual human specifically, it generates a profound effect on their inclinations, behavior, and action, eventually gripping them in a vice, similar to a cog in a machine. The predicament is, common people were not aware of or chose not to find out the output of this machine, and if they were lucky enough to see the yield, it usually is pouring into the interest of a King, a High priest, an individual human or a group thereof. It just doesn't feel right!

But the Era of the horse is no more, and now we have fast electric cars. Emperors became merely decorative, and their authority is replaced by "Multinational Corporations" or "Stakeholders."

The irony is the new rulers of the world are doing what the old ones have done, looking for a collective consciousness that unite their subjects. "Materialism" seems to serve the purpose very well, as it makes their merchandise the only measurable element in the human world. But this has set them on a conflicting path with the idea of theism, which by default is anti-materialistic. Thus, through raising the flag of the freedom for the human race with unprecedented support for the evolutionary theory and atheism, they have hoisted the foundation of the New World Temple.

While the old Emperors told their subjects they need to avoid materialism and sacrifice part of their produce for the benefit of the King. New Emperors are imparting to their subjects to consume more, work twice as hard, and abandon theology.

Both old and new state of affairs are various forms of human slavery. The truth should not allow any one human control over another, and this is why we should care because if we don't, we will be exploited by one group or another.

Hypothetical: World scientists and cosmologists have agreed the universe is an intelligent design with a colossal amount of energy; we already know this energy constitutes 32% of the total energy, and the rest is lying within what is termed dead space and dubbed as dark energy, this is a repulsive force that helps the universe to expand. The amount of power and knowledge we are contemplating here is gargantuan and can have serious consequences on us.

The obvious notion is we, "the human race," are so infinitesimal in size and power compared to our surroundings. We have seen from our previous arguments and deductions how we are dependent on the Design process and, in turn, upon the Designer. Audacious claims of abrogating the owner of this design sounds either risky or foolish.

We must attempt, for our safety and future, to find the truth. We can start by looking for a message that might have been relayed to us.

Translating this to a dramaturgical scenario, let us envision that the United Nations of the World has, through very sophisticated technology, suspected a gathering of a massive power of forces at the border of our galaxy; they decided to monitor the situation and the conduct of this armada. Time goes by, and nothing happens. Scientists and experts start to scrutinize their behavior and disposition, finally concluding this mighty force does exist and is overpowering and knowledgeable.

The debate in the top brass meeting of the council is as follows: Whether this force has done nothing to us so far, why should we care? Probably carry on, business as usual. These guys are in conflict with another group, who are worried about the future, claiming so much power and knowledge if exists, can potentially have dire outcomes on us all. Thus, we must attempt to find out more and probably look for a message or a messenger these powerful armies might have sent. Ultimately, we can't afford to ignore that much power and knowledge and the only way to secure our safety is to make peace.

SCIENTISTS VS POPULATIONS

Atheists have repeatedly used statistical figures demonstrating a lower proportion of theological belief among scientists compared to the normal population,[165] in an attempt to support their deliberations. Inferring, scientists must be more intelligent. Referring to a survey conducted by the Pew Research Centre in 2006, which showed 83% of Americans say they believe in God. Edward Larson, a historian of science at the University of Georgia, USA, did a poll in 1996 and came up with a finding that 40% of scientists believed in a God, while 45% said they did not.

The logic behind the atheistic claim is totally skewed and here is why:

Intuitive Thinking: Research work in business psychology and intuition has been conducted by Daniel Kahneman, a Nobel laureate of 2002 in economics, who described intuitive thinking as a function of an area in the brain which acts very fast without thinking anything through.[166] This is complemented by a slower system that comes to a more precise decision after checking the reaction of the very fast system. These two systems are interdependent and work together.[167] They decide in an extremely short time. This allows businessmen to make intuitive decision in situations they deal with daily.

---

[165] https://www.pewresearch.org/religion/2009/11/05/scientists-and-belief/
[166] Intuition as a Basis for Business Decisions. Prof. Dr. C. Schmidkonz, Patrick Stütz. 2017.
https://www.munich-business-school.de/insights/en/2017/intuition-business-decisions/
[167] Thinking, Fast and Slow. April, 2013. by Daniel Kahneman.
Farrar, Straus and Giroux; 1st edition. ISBN-10 : 0374533555

Researchers at Harvard University, USA, found people with more intuitive thinking styles tend to have stronger beliefs in God and religion. Does this make them less intelligent than scientists who tend to use analytical thinking? In this study, the researchers tried to test the assumption that belief in God is a natural product of the human mind. They questioned this assumption by saying if this is true, then they should find belief in God to be influenced by one's tendency to rely on intuition versus reflection. To test this supposition, they conducted three studies that supported this hypothesis, linking intuitive cognitive style to belief in God. They added: *"This effect was not mediated by education level, income, political orientation, or other demographic variables."*

Analytical Thinking: A critical style of thinking, with its basis in logic, rationality, and synthesis, relies on hard facts and is what enables new discoveries and innovations. This is believed to be the hard-core thinking skills needed in undergraduate and postgraduate teaching of science.[168]

In a paper published by the Journal "Science:" "Analytic Thinking Promotes Religious Disbelief." by W. Gervais and A. Norenzayan. They examined the hypothesis stating: *If religious belief emerges through an intuitive process, and as analytic processing is known to inhibit or override intuitive processing, then analytic thinking may undermine intuitive support for religious belief.* They concluded: *Subtle manipulations known to trigger analytic processing encouraged religious disbelief;* they also warned: *the findings do not speak directly to conversations about the inherent rationality, value, or truth of*

---

[168] Approaches to Learning and Teaching Science. Mark Winterbottom, James de Winter ISBN:9781316645857.

religious beliefs; they illuminate one cognitive factor that may influence such discussions.[169]

My understanding from all of the above-mentioned research is using a certain type of cognitive processing is likely to influence your belief, as we have agreed previously every human should have a multitude of thinking skills acquired as a result of life experience and training, with most humans having a dominant type they are comfortable to exercise most of the time. In an ideal situation, we should be equipped with more than one cognitive and thinking aptitude in order to deploy the most suitable skill to the appropriate task; otherwise, we will end up in a cognitive bias or our views in life and career will be skewed towards one direction or another.

To link these studies to the atheistic claim, both intuitive and analytical thinking are great tools of our brain, except when we apply them to the wrong issues. An example besides that we mentioned in Chapter II is when a crime scene investigator uses critical thinking and analytical skills to analyze a crime incident, then he uses the same thinking skill to organize his marital life.

In the matters of faith and belief, if you attempt to utilize analytical thinking to decide about the correct faith or the correct message from our Designer, you are likely to decipher right from wrong, but if you apply the same skill to issues of consciousness limitations, we mentioned in Chapter I, you will only reach a dead end, then you will be used as statistics in the atheistic claim.

We can conclude human beings have different styles of thinking skills, but this does not mean one individual is more intelligent

---

[169] Gervais WM, Norenzayan A. Analytic thinking promotes religious disbelief. Science. 2012 Apr; 336(6080):493-6. PMID: 22539725.

than another; a very successful businessman cannot be called less intelligent just because he tends to use intuitive thinking skills compared to a scientist who tends to rely on analytical thinking. But it is the application of the wrong thinking skill to the task needed that can make people take certain views of belief different from each other.

THE EXTRA-TERRESTRIAL BIAS

It seems so ambiguous, thus illogical, that many humans are fixated on the idea of looking for a message from extra-terrestrials. I was also surprised doing some research on this topic that many atheists are almost certain of the presence of life on other planets. Solely because there are many stars in the universe and extrasolar planets are ubiquitous. I am not against the possibility, but I find the above two inferences totally inadequate as evidence. Logically, whoever created humans could have easily created Bumans or even Cumans. But I cannot fathom how somebody says I cannot believe in a God (something I can't see) becomes so certain on another similar topic, life on another planet, despite having some evidence of the first occurrence, which is in the least, his own existence, and without any proof of the latter.

This goes back to the same false premise: to believe until evidence shows up. The least you can describe this is it is against basic scientific methodology. Or just "Bias," as the presence of multiple scattered life could give the impression of an unintentional self-creation. In contrast, this whole universe with only one live-able planet gives an impression of an intentional creation. But in argumentized logic, neither statements are necessarily true.

There is no evidence the laws of physics at this end of the universe are any different from those laws of physics at the other end of the universe,[170] except for the wrapping of time. (The faster you travel through space, the slower you travel through time, and is dubbed as time dilatation).[171] While humans can never travel more than the speed of light as long as they have mass, if they lose mass (body), their soul may do that, but this is something we will all find out later when we lose our mass (body).

So, we can conclude, given the vast distances involved, it is doubtful we will ever encounter another species as long as we are incarcerated by the four physical forces of the current universe. But this does not mean in the least extra-terrestrials do or don't exist, only we will never get to meet them in our current form.

---

[170] This is now contested by physicists who believe in the Multiverse, or an expandable universe with pockets of universes that is governed by different laws. However, the 4 physical forces must still work, as they cannot infer that atoms will not have electrons.

[171] https://en.wikipedia.org/wiki/Time_dilation

VII

The Message

THIS IS A CORRECT MESSAGE
The Designers' Brush Stroke

Suppose you imagine a panel of artists standing in front of a newly discovered painting canvas. The usual questions would be: How old? and who is the artist? If this information were not written clearly on the canvas, they would go into a scientific carbon dating process to check the age, call experts on the most similar technique, and shortlist artists names who lived during that era. They will also look for hidden messages within the canvas which might give away the master of the work.

Returning to our point of discussion, "Design," and imagine a second scenario where two large canvases are displayed in an art museum, where an artist decided to illustrate the design of the smallest and the largest units of our universe side by side. The first is the Atom, and the latter is the Galaxy; both are units with trillion duplicates. The noticeable part will be, both would look as if they have the same design technique. The first is the nucleus, with the electron clouds rotating around it in concentric orbits. The latter is a sun with the planets revolving around it in concentric orbits. Both have the same brush stroke.

In a design, when the smallest unit has the same technical concept as the largest unit, a logical argument could easily infer they belong to "One Designer."

Logical reasoning should make us go looking for any hidden signatures, messages, or marks which might lead us to confirm who the Originator is.

As we know by now, there is no clear signature on our universe's canvas that can be read out loud, but our current scientific knowledge will allow us to compare notes with all messages humans might have received from the beginning of creation. I am certain we can then verify and authenticate a correct message, mainly by comparing the communication content with the canvas design.

What Constitutes a Compelling Message
from an Intelligent Designer?

It is unreasonable to believe that a designer of such a complex universe would not send a message to guide his human creation, or leave them clues about how it was made, even if those clues are hidden and require some search or effort to unearth. This is consistent with the "Examination hall theory," i.e. we have a chore to learn about the universe by looking for valid designer messages. Once we find it, will be able to answer many important uncertainties, such as why we were created, where we are heading, and what constitutes good and evil.

In our quest to search for messages from the Designer, we must be as systematic and logical as possible. This means examining and checking every single claimed message, even though the claims are endless and new ones appear all the time. A more noble approach could be to look and decide about what to expect from an ideal message. What it should or should not be; this will shrink the list substantially.

Pre-Condition Your Brain: Before starting this mental exercise, it's essential to put your brain in a pairing mood to percieve the message. The metaphor here is you need first to cancel your bias; second, you need to inhibit your normal neuronal pathway, which tends to make you think along certain paths, thus inhibiting all other possible pathways (same as Newton did, which allowed him to see the concept of gravity by noticing the apple falling).

To help the reader meet the above requirements, I will highlight a few issues which might influence his perception of logic.

- *The Followers:* This is a predicament that can skew our deliberation, which is the difference between the message and the adherents of the message. An example is some believers tend to aggregate in a sectarian fashion according to colour or race, but their message never said that; thus, one should not ascribe the follower's behaviour to the legacy of the message.

- *Movie Logic:* Although the movie industry started in the 19th century, it wasn't until the 20th century when the spread of films began. According to the Cornell University website, it was between 1950 to the 1970s when television exploded at a greater than exponential growth rate. This means any human living today below the age of 70 would have been raised and influenced by the film industry, particularly its stronger arm "Hollywood."

  Why is this point relevant here? It's the expectation and setting of society's standards of right or wrong, what is and what should not be acceptable, and most accurately, good and evil. With its financial and far-reaching power, this human-invented industry has not only penetrated every house on planet Earth but tried and is still trying to change our perception of what is good and what is bad. Try, for example, looking at productions related to Islamophobia or Homosexuality and check which direction they are trying to take you. Here, I am not siding with or against any group but demonstrating that the movie industry does.

Now to my theory of Movie logic: Since I was a teenager, when I started developing a fascination with movies like everybody else, I had a very strange and consistent observation. This was neither the obvious, that there is always a good guy and an evil one, nor that the good guy never dies and always wins, as they started changing this rhetoric a little later. But it's the constant finding that the narration, speech, and aim of the evil guy in the movie industry is usually a copy and paste from religious text and ideology of various faiths. It is tough to prove this impression in this context, but I would like the reader, for now, to temporarily suppress the movie logic that has built up in his mind over the last 70 years, and maybe later, he can check the validity of the above statement, retrospectively or prospectively.[172]

- *Religion Shopping Centre:* It is essential to remind ourselves we are here to find the truth, not something good. As good or less good are social observations, they can mean different things to different people, but the truth is one.

To be clearer I have met fellow humans who said they have

---

[172] Try looking at the film "The Island" (2005), which I enjoyed watching, and as a mental exercise, look for the theological analogy the film is imparting.

investigated several books and messages and found this or that to be satisfying. When they quote the reason, it typically sounds like they were shopping for a new car. Have they not asked themselves what evidence they were given for the message's validity and whether the promises given to them can be delivered or not?

My most common encounter is fascination with minimalist or frugal living. If the Designer didn't make all the comfort on the planet for humans to use, then for whom.

- *Modernity vs. Antiquity:* I have read and heard numerous arguments about science and faith, where authors agreeing on the principle of Intelligent Design consistently warn about the notion of going back to the past quoting specifically bad historical times like the Middle Ages, more specifically, Inquisitions, Holocaust, and Crusades. Although I conquer these were some of the worst events in human history, we must remember they were man-made and were another example of human unfairness we discussed earlier.

A message from the Designer, if exists, must have happened in the past, maybe with different versions. The mind trap here is linking the bad history of antiquity to a claimed message just because some humans alleged their God told them to do bad things.

The Ideal Designer's Message Characteristics

Declaration of Self: It is logical that a Designer of a universe with such complexity to commence his message with a declaration of self, in an unequivocal and unambiguous statement

Declaration of Intent: One would expect a declaration of intent of what is to be of the human race as a whole.

Validity: We have to presume our Designer has full knowledge of our tendency to sacrifice truth for profit. We can easily claim that the history of humanity is full of mythical stories of proclaimed Devine Power, acquired through heritage, lineage or connection; its end result was mainly aimed at consolidating power in the hands of humans, whether Kings, Generals or High priests. And they achieved this by subduing other humans into financial or physical labor. It is also fair to say human deceit has probably tried and will constantly attempt to hijack a valid message for the same benefit.

Thus, the "Validity" of a message is a must to help humans decipher a genuine from a fake message.

A historical story of what some people call a miracle does not qualify as validity or evidence of any message. As we were not a witness to it, the time-lapse means it is impossible to verify the events historically. Thus, a story of a miracle should remain as it is, just a story.

Timeline: We should assume the Designer would have decided to send his message from the beginning of creation of the universe. Thus, we expect his message to cover this time span by explaining or following up on previous messages. Messages popping up in the middle of history favoring one group of humans over another and not linking up with previous human history is probably a human work aimed at making their author popular.

We also would expect this message not to be time-bound to events of its inception.

Logic: It is unreasonable to expect a message from a designer with such vast knowledge and power to be illogical or to forbid humans from using their brains. If one hears or reads the phrase: You can't use logic, one should run and never look back.

Demarcation of Good and Evil: Our early attempts to create our own standards for Good and Evil resulted in mediocre laws. While we have made significant progress, we still struggle to define Good and Evil clearly. Thus I believe humanity needs some principles to follow.

Equality: As we have demonstrated earlier, how the natural behavior of developing eight years old children would suddenly favor fairness and equality; thus, the message should do the same as it would confirm with the Designer's technique we see on the ground.

Financial Structure: The presence of a functional economic pyramid, with a human sitting on top who have access to funds, is definitely a no-go scenario, as financial structures tend to defend their own financial interests rather than looking into matters at hand.

Hierarchical Leadership or Non-hierarchical Structure: It is natural for humans to work under some leadership. A message spanning the whole of humanity should not include everybody in a rigid leadership structural pattern. The reason is the design feature of diversity will be lost. How, then, can we expect a message from a Designer who created diversity to allow a structure which would eliminate it?

Also, hierarchical Leadership and non-hierarchical structures, if successful, one can attribute their success to the team working on them which should not be the case. Another bad combination is when leadership claims divinity or speaks on behalf of divinity. It just doesn't feel right.

Science and the Message: Science is the study of the natural world, and the latter is the product of intelligent design. If a message from its proprietor was deficient in science, it would be like a message from a builder that did not contain any information about construction.

A message from an intelligent designer would neither be too simple nor too complex. Meaning, complex enough to contain valuable information, but not so that it would be incomprehensible. It needs to be compatible with our current understanding of science, but not limited by it. Thus I think an optimum balance of complexity and simplicity enough to be challenging and informative, but not impossible to understand or overwhelming. I expect it to be compatible with our current knowledge, and may be expand our understanding of the world.

One or More: Modern humans have existed for around 60,000 to 80,000 years. Scientists have used genetic markers and ancient geography to partially reconstruct how humans migrated around the world. It is believed that the first humans to explore Eurasia used the Bab-al-Mandab Strait, which now divides Yemen and Djibouti. They reached India by 50,000 years ago, and then Southeast Asia and Australia.[173][174] It would make sense for a designer of the human

---

[173] https://www.universetoday.com/38125/how-long-have-humans-been-on-earth/
An article by E. Howell 2015 on Universetoday.com

She writes: The modern form of humans only evolved about 200,000 years ago. It is presumed here that she is referring to the Homo Sapiens, who were an erect form of humanoids, that fossil evidence have shown their presence around this time.

[174] According to the Encyclopaedia Britannica, the oldest remains attributed to H. sapiens were found at a site in Ethiopia's Omo Valley. According to the same source, a newer discovery in 2017, these dates have been challenged by an excavation by Jean-Jacques Hublin, of the Max Planck Institute for Evolutionary Anthropology in Leipzig, Germany, which revealed that H. sapiens was present at Jebel Irhoud, Morocco, The team unearthed a collection of specimens which were dated to about 315,000 years ago.
https://www.britannica.com/story/just-how-old-is-homo-sapiens

race to send a series of messages, rather than just one. However, these messages must be consistent with each other, otherwise we would need to explain why the interest in some particular epochs of human history and not the other.

Guide to the Meta-Universe: We have agreed about the merciful nature of the Designer; it would suit this character if he gave us in his message a guide about the path to the meta-universe and what to expect in the after-life.

Good Deeds: All sects of beliefs recommend good deeds, but we should look between the lines. To be honest, "Love thy neighbor" sounds a bit lame or could be more of a campaign slogan than a knowledgeable designer message. What if my neighbor is a serial killer. Absolute love and peace are virtuous in theory, but they dont have enough sophistication to match our real-life scenarios or our designer's complex work.

What do I expect? Answer: A system of fairness which promotes a peaceful relationship between humans and human groups, yet prohibitive and even proactive against wrongdoing. So, you don't get to turn the other cheek.

Bidirectional Communication Channel

When I bought my beloved car a few years ago, I was told the mother company could communicate with the car, send messages for service alerts to my dashboard, according to the readings the computer onboard sends to them, i.e., they will decide on the suitable time to do the service based on the readings from monitors built-in, like oil parameters and other stuff. I am sure you will believe me if I say they are on a different continent than the one I am living in. Also, the car automatically uploads data of any major events that can happen during driving to the manufacturer, similar to an airplane black box, for analysis later if necessary. I was also told, this at the moment, is confidential data. Consumer Reports magazine estimated 32 of 44 car brands offered some form of wireless data connection in their 2018 model-year cars. By the year 2030, all cars on the road could be equipped with such data transmission systems.[175]

This two-way communication in a car design was easily achievable by human designers, and we will probably see more advanced forms in the future.

Deducting car design is probably way less sophisticated than human design, could the same phenomena or a two-way communication device be present in our current human form? Are there any premises to infer such an idea?

The deeper I delve into this thought, the stronger is my conviction that a bidirectional communication mechanism must be in place. Yet, where's the evidence?

Many people throughout history have claimed to have communicated with a divine power. While it is possible that some of these people were lying for personal gain, have we scientifically examined these claims and subjected them to logical scrutiny before concluding they are false?

Throughout history, the concept of a creator has been self-evident, and humans have always claimed a God figure out of literally everything around them, including animals and statues. Logically,

---

[175] https://rollcall.com/2019/04/09/your-car-is-watching-you-who-owns-the-data/

we can presume the communication port in this situation was opened on the human side but was not receiving any legible data from the animals or statues. But out of the few thousand humans who claimed Devine contact, could there have been a message with legible data communicated between the Designer and a human subject. First, we must examine the available data set and subject it to laws of science and logic to find the evidence. We have already discussed the required parameters we should ideally expect in this data set or message.

Scientists have detected the emission of brain waves from our heads, which are detectable and measurable. Still, they can identify the different states of consciousness the person is in or which part of the brain is being activated. Our brainwaves change according to what we're doing and feeling. Although science can not read what a human brain is thinking in details yet, but it can detect which part of the brain is activated. And is probably a matter of time before scientists will be able to know more. This already proves a one-way communication, i.e., we send signals outside our skull; the second query is, who is listening? One can inductively claim or infer from the car model, the manufacturer or the Designer can.

Hypothetically, every human can attempt to communicate with their Designer, which might activate the communication port. He can see for himself if it is possible to receive guidance directly. My advice in this situation is you don't want to send a message like "Hi, if you are there, can I see some fireworks," because either your communication port will be shut down, or your wish might come true. In both situations, you are not the winner. Instead, one should ask humbly for guidance.

SUBITIS CREDERE (EMERGENCY BELIEF)

How much time do we have to make up our minds? The answer is I don't know; I presume when the bell rings and it is time to go, that's when the exam process will end and the marking starts. So, if we use the current logic and known metrics, we have till the end of our personal life. Thus, if you are in your twenties, you probably have ample time to find out. Otherwise, you will need Subitis Credere (Emergency Belief).

In this philosophy, one can use retrograde logic instead of checking for every possible claimed message and then verifying its authenticity, whether it is from the universe Designer or a fellow human quack. You can start by looking at the final product, "The Design," and start believing there is no God except the God who created your human form, fellow humans, and the existing universe.

I can't see how anybody could disagree with this logic unless one has a preconditional bias towards the principle of being designed. Or doesn't perceive the prospect of being told what one should or should not do.

Then, you can hope you have enough time left to find the correct message.

The Right Message

If I claim which message, I believe is the right one. This will almost certainly alienate the majority of potential readers. The human mind doesn't like to be told what to believe. I would instead leave every reader to his own conviction so he can envisage the logical trail and arrive at his own supposition. Besides, I would have been accused of bias anyway. Using logic to dissect facts would be a more invigorating approach, and the reader has the choice to believe in what he desires.

It's evident the Designer has allowed humans space and time to think, contemplate, appreciate design, and then arrive at their own assumptions and beliefs. How could I then come to push it into the reader's mind?

Intelligent Design Revisited

By enlarge, the majority of thinking groups believe in the concept of intelligent design, or at least cannot argue against it. However, there remains skepticism over the notion, an Intelligent Design can have a Designer, which doesn't seem reasonable. Their explanation is always the presence of "pain, injustice, disease, and suffering." We have discussed these parameters; who is causing them? And what could be the reason behind incorporating them in an Intelligent Design?

If we revisit the term "Intelligent Design," we need to be more articulate and incorporate other design features. A deed, which on occasions, include the presence of "pain, injustice, disease, and suffering" combined with intelligent parameters.

The description "Incomplete" we have discussed earlier not only complements the logical narration we have used throughout the book but also gives hope to all humanity, whoever endured pain, injustice, disease, or suffering would not have done so in vain, but would have been thoroughly accounted for, will be compensated or venged. The balance and symmetry we have observed within our human form and the universe's architecture will not stop short of humans and the human race's behavior or their deeds.

Thus, I am hypothesizing the use of the term "Intelligent Ad-Interim Design."

Can Science Lead to Faith?

Coming back to address our opening question: Can science lead to faith?

The question implies science comes first, but in reality, and for many people it doesn't.

The Answer could be: Science is good at explaining "How things work" but not necessarily for explaining "Why do such things exist?" or "For what purpose?" Science does not really explain why the universe exists, but it can explain how it functions, or at least is trying to.

Science, with its four physical forces, does not explain everything in our universe; there are other significant parts of our human existence that are non-science bound (our Soul, maybe our Consciousness and maybe many more things that we can't realize or perceive).

Science has a very narrow vision and can only look at evidence through a microscope, but with evidence, one can verify faith.

Humans should not be solicited to have faith without evidence. Lack of proof polluted the image of religion in people's minds, as humans started to claim so many things which are baseless, resulting in two scenarios, one creating disdain out of the idea of Deism, and two loss of faith.

Science serves as a key for looking, assessing, appreciating, deciding, and believing in the message from our Designer, thus contributing to faith but not finding it.

When some faiths perceived science as a threat and a destabilizing force, they attempted initially to subjugate it, then later quash it. As science grew stronger, they started losing believers.

When science allowed humans to reach the moon, cure diseases, and impact their living environment, they started developing respect and admiration for its processes. But when they fail to recognize its limitations, they begin setting it on a conflicting path with faith.

Scientists, on the other hand, thought by adopting the evolution theory, they were scoring a goal in the faith side and they had rescued humanity from the grip of medieval theologians. The problem is, any fair referee will dismiss this goal as an offside kick. The reason is the hypothesis proposed by Darwin, albeit genius and novel for its time, never reached the scientific level of theory, as the evidence available so far does not support its validity. Yet, for the first time in the history of humanity, scientists have called an idea a theory without having proved its premises as factual. Instead, it could be labelled as a hypothesis or an assumption.

Science and theology are complicated labyrinths for the human brain to navigate, even if this brain is intelligent or well-read. We can observe many examples of top intellectuals going into these mazes and never coming out with definitive conclusions. In other words, they still have many unanswered questions. The only link between science and faith must be a message from the designer of science, who, by default, is the owner of faith.

Humanity, with its intellectual powers, has been wired to have many capabilities, but many others are beyond our conscious sway.

No matter how hard we try, we fail. Like grasping any idea about our soul, which lies within ourselves, how much closer can this be? Yet, how unfathomable it is to comprehend.

Avid scientists declared a conjecture that science can explain all, ignoring the possibility that anatomically, our consciousness has not been programmed to feel or comprehend many things. But they are still promising their followers with potential hope, reminding them of their little successes in natural science.

Avid theologians have been less fortunate, as their narration, by default, comes from antiquity, which has been linked in the mind of humanity with misery and abuse, which they have helped to create. To compensate for their failure to defend faith, they used tactics ranging from concessions to fundamentalism. Both have failed miserably, as seen from the exodus we observe.

The bridge between the two itinerant webs of Science and Faith can never be built without a manual, catalogue, guide, or message from the designer of science and owner of faith. Without a genuine guiding message, both islands (science and faith) will drift at their own momentum and can hardly meet.

One thing is quite certain: the creator of the physical forces (science) has given us the tools to comprehend his intelligent work for a reason. And that is mainly to find faith and not food. The latter could have been readily and endlessly available, as seen in other sections of the design. We share this universe with many animals who don't possess our intellectual powers, yet they are well-fed.

About The Author

I am a practicing Oral and Maxillofacial Surgeon with a main interest in the reconstruction of traumatic facial injuries and congenital facial deformities. I had my undergraduate education in Cairo, Egypt, with postgraduate education and experience in the UK. And surgical fellowship and membership from the Royal College of Surgeons of Edinburgh and Glasgow. I have done rotational training in other countries, Germany, India and where I practice now in the United Arab Emirates.

I am married with three children, who are professional and successful, and this alone makes me incredibly proud and satisfied with my life achievements. My wife is a very successful children's book author who published many Arabic books promoting the education and spread of the Arabic language, especially for school children.

I believe in God as the creator of our universe and follow a mainstream religion, and I prefer to leave it at that.

I hate being categorized into a sect or subgroup and prefer not to delve into this path. While I've consistently refrained from criticizing these groups or denominations, this stance might warrant some clarification.

Documenting my logic behind not labelling myself under a sect or group might serve as a logical guide to others. As I often find all sub-sects or groups are by default classified under a human founder, which first doesn't make sense; second, with the passage of time, this position often gets inherited genetically or hierarchically acquired by another human, who instead of continuing on the original path, can decide to drift to the right or to the left. The effect of this swing on people like me and you, who usually sits at the bottom of the pyramid in the hierarchical scale, is suddenly finding themself at cross confrontation with society, other sects or their original beliefs. In short, I am not fond of much human influence or guardianship over beliefs.

Suggested Reading

- There Is ~~No~~ A God: How the World's Most Notorious Atheist Changed His Mind. Paperback. 15 November 2008, by Antony Flew (Author) and Roy Abraham Varghese.

- More than Myth? Seeking the Full Truth about Genesis, Creation, and Evolution, Chartwell Press, 2014. By Dr P. Brown & Dr R. Stackpole. In a chapter written by Casey Luskin, "The Top Ten Scientific Problems with Biological and Chemical Evolution," He puts a very logical scientific analysis of why biochemistry or biology does not support the evolutionary theory.

- Lateral Thinking: A Textbook of Creativity, by Edward De Bono, Published 1991 by Penguin UK.

- Darwin's Doubt: The Explosive Origin of Animal Life and the Case for Intelligent Design. By Stephen C. Meyer, June 2014.

- Poverty and Famines: An Essay on Entitlement and Deprivation Paperback, January 1983, by Amartya Sen FBA.

- The Devil's Delusion: Atheism and its Scientific Pretensions. By David Berlinski September, 2009.

- The Deniable Darwin, by David Berlinski. Kindle Edition (ISBN 0979014123).

- The Biology Of Belief: Unleashing The Power Of Consciousness, Matter & Miracles, by Bruce H Lipton. 2010. Hay House Inc. ISBN-10 : 140195247X.

- The Inner Light Theory of Consciousness, February 2, 2002. By Steven W. Smith.

- On Writing Well: The Classic Guide To Writing Non Fiction. 2016, By William Zinsser. HarperCollins Publishers Inc. ISBN10 0060891548.

- Eye and Brain: the psychology of seeing. Fifth edition, 1977, by Richard Gregory. Princeton University Press, ISBN-10: 0-691-04837-1.

- The Believing Brain: From Spiritual Faiths to Political Convictions: How We Construct Beliefs and Reinforce Them as Truth. By: Michael Shermer. Publisher: Hachette, UK 2012, ISBN 178033530X, 9781780335308.